Dangerous Earth

Dangerous
EARTH

WHAT WE WISH WE KNEW ABOUT VOLCANOES, HURRICANES, CLIMATE CHANGE, EARTHQUAKES, AND MORE

Ellen Prager

The University of Chicago Press

Chicago and London

The University of Chicago Press, Chicago 60637
The University of Chicago Press, Ltd., London
© 2020 by Ellen Prager
Published 2020
Printed in the United States of America

29 28 27 26 25 24 23 22 21 20   1 2 3 4 5

ISBN-13: 978-0-226-54169-3 (cloth)
ISBN-13: 978-0-226-54172-3 (e-book)
DOI: https://doi.org/10.7208/chicago/9780226541723.001.0001

Library of Congress Cataloging-in-Publication Data

Names: Prager, Ellen J., author.
Title: Dangerous Earth : what we wish we knew about volcanoes, hurricanes,
    climate change, earthquakes, and more / Ellen Prager.
Description: Chicago : University of Chicago Press, 2020. | Includes biblio-
    graphical references.
Identifiers: LCCN 2019025458 | ISBN 9780226541693 (cloth) | ISBN
    9780226541723 (ebook)
Subjects: LCSH: Natural disasters. | Climatic changes. | Hazard mitigation.
Classification: LCC GB5014 .P73 2020 | DDC 551—dc23
LC record available at https://lccn.loc.gov/2019025458

♾ This paper meets the requirements of ANSI/NISO Z39.48-1992
(Permanence of Paper).

*For Dave, whose passion for family and
weather knows few bounds,
for bringing so much joy, laughter,
and love into my life*

Contents

Earthly Dangers and Science

The Earth is a beautiful and wondrous planet, but also frustratingly complex and at times violent. Much of what has made it livable can also cause catastrophe. Volcanic eruptions create land and produce fertile, nutrient-rich soil but can also bury forests, fields, or entire towns under ash, mud, lava, and debris. The very forces that create and recycle Earth's crust also spawn destructive earthquakes and tsunamis. Water and wind bring and spread life, but in hurricanes they can leave devastation in their wake. And while it is the planet's warmth that enables life to thrive, rapidly increasing temperatures cause sea level to rise and weather events to become more extreme. On Mother Earth, it is a love-hate relationship for the planet's residents.

Humans have been dealing with the dangers of living on Earth since our species first arose. In years past, calamities were often explained through myth or religion. Today, we look to science for answers after disasters strike.

But science isn't a static process. It involves observing, testing, and retesting. It is a process replete with failure as well as success and controversy. Some ideas yield new insights that become part of our fundamental understanding of how the planet works.

Other concepts go down the proverbial toilet. Over time, it is the process of doing science that adds to our growing knowledge about the Earth. And along the way, new questions continually arise, while mysterious unknowns may linger.

The unknowns are also what drive science. They fascinate and frustrate us. They ignite our curiosity and passion to learn. When we think of the unknown, we often imagine the strange, the alien—things beyond our planet, our reach, our everyday life. What lies in a galaxy far, far away? What strange creatures inhabit the ocean's unexplored depths? But all around us are amazing mysteries: things that remain undiscovered, unresolved, and in some cases unimagined—the unknown unknowns.

Science is critical to society, and it is constantly evolving. Science is about what we know, what we don't know, and how we learn. This book is not meant to be a comprehensive guide to the phenomenon described. Its goal is, rather, to highlight game-changing events—some tragic, yet astonishing—and to consider what was learned from them, and what remains unknown.

When it comes to planet Earth and its most powerful forces—forces that have led to disaster in the past and will do so again in the future—the scientific unknowns are not abstract, esoteric topics. They are the wellspring of questions whose answers could help prevent tragic and catastrophic loss. Why can't we predict earthquakes and tsunamis, or when and how a volcano will erupt? Why can't we forecast where and when the next giant landslide or mudflow will occur? And what makes it so difficult to determine, in a warming climate, how fast and how high sea level will rise?

Part of the answer to these questions is that the phenomena involved are frustratingly difficult to study. Many of the Earth's processes are dynamic, ephemeral, and their origins are hidden from view. Humans are short-lived creatures on the skin of the planet at the base of the atmosphere. We cannot readily see into the ocean's depths, beneath Earth's rocky surface, below polar ice, or above the clouds. Our lives, indeed, our historical record of events, are a blip in the planet's billions of years of existence. The most powerful and destructive events are also fortunately rare, but their infrequency is an obstacle to learning. Like the unknown, these conditions challenge scientists, yet also inspire them to persevere and develop new and innovative means of studying the Earth, the ocean, and the atmosphere.

Scientists now use sophisticated and rapidly improving technology to learn ever more about the planet, pursuing research even in extreme conditions and previously inaccessible environments. They have learned to expect the unexpected—it's how science works, evolving as new data are obtained and give birth to new ideas. Data are critical, and collecting more is the goal. But how much data or knowledge is needed before theories become fact or action can be taken? Do we know enough now, even with unknowns remaining, to reduce risk and save lives and property? Sometimes it's what we already know that keeps us up at night.

As human populations continue to increase and spread across the planet, more people than ever are at risk, from natural disasters and from crises humans have created or exacerbated. We've learned a lot about how the Earth works, about the powerful forces that can cause destruction, but many mysteries remain.

In the following pages, I've recounted some of the astounding and often tragically destructive events that have changed our worldview. I consider what was surprising about them, what science was learned from them, and—just as important—what scientists still wish they knew. For scientists, it is not a weakness to say, "I don't know," but a strength. Because admitting we don't know opens the door to exploration and inquiry, which are fundamental to learning.

Science can sometimes be confusing, especially the technical details. So I've tried to keep it simple, to explain some of the basics and use compelling stories, fascinating projects, and vivid images to make the science more user-friendly and engaging. My apologies to the many programs, projects, organizations, and scientists I've neglected to mention. Some who have generously spoken or corresponded with me appear in the text; others are acknowledged at the back of the book or can be found in the list of readings and references. In the chapters themselves, I've kept the acronyms, jargon, and who-did-what-when to a minimum and focused on, again, what we know and, even more so, what we wish we knew. The information provided is based on data, not ideology or beliefs, except where stated.

I hope you enjoy the book and come away as fascinated as I am, both with what scientists have learned about the phenomena described and with what they still wish they knew. And I hope you will be moved to support the researchers, policy makers, and others working to make life safer for all of us who call Earth home.

1

Climate Change

Antarctica is a sleeping elephant that is starting to stir. When
Antarctica fully wakens, it will likely be in a very bad mood. –**Mark
Serreze**, director, National Snow and Ice Data Center

WEST ANTARCTICA, SUMMER 2002. It was unusually warm. But
not all that unusual—over the past fifty years, summers in West
Antarctica had become some 2.5°C warmer. The last several had
been particularly balmy, and now strong mountain winds amplified the higher temperatures. Across the region's slow-flowing
rivers of ice, or glaciers, meltwater trickled down through holes
and cracks. As water percolated downward, the cracks grew wider
and deeper. Soon meltwater reached the base of the glaciers. Normally, friction from the underlying bedrock slowed the delivery
of ice from the vast West Antarctic ice sheet to the sea. But now
as meltwater reached the base of the glaciers, it lubricated the
flow and allowed the massive rivers of ice to march seaward more
speedily.

At land's end, the glaciers pushed beyond the underlying rock
to form massive ice shelves that floated buoyantly over the cold

Figure 1.1. The leading edge of the Larsen B Ice Shelf as of 2008. HD/Reuters.

polar sea and blocked the flow of the glaciers behind (figure 1.1).
In the summer of 2002, one such ice shelf, thousands of kilometers wide and 220 meters thick, lay off Antarctica's horn-shaped peninsula, adjacent to the Weddell Sea. It was a section of the Larsen Ice Shelf known as Larsen B (Larsen A lay to the north, Larsen C to the south). Summers of melting and winters of freezing had created a slick sheet of ice on the surface of the Larsen B Ice Shelf. By February, late summer in the Southern Hemisphere, small ponds of meltwater lay like dotted lines along sutures between old glaciers whose slow flows had long ago coalesced.

As the warm days continued, the meltwater ponds grew. Water seeping down into the sutures caused fractures to form and deepen. Meanwhile, at the base of the shelf, relatively warm ocean water lapped against the ice, carving channels in its underside, and weakening it. Soon, within weeks, the massive Larsen B Ice Shelf began to splinter. As giant slabs and tall, narrow strips of ice broke off the shelf or calved, thunderous whumpfs echoed across the region. Wind, waves, and currents tossed and tipped the blocks of ice. Some of the newly born icebergs clustered like

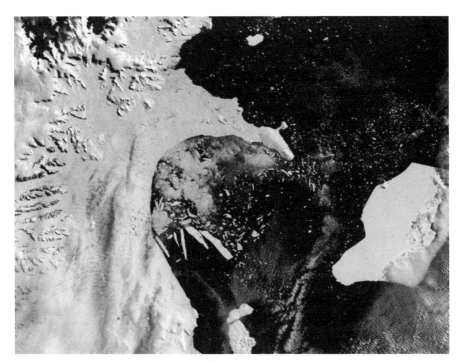

Figure 1.2. Larsen B Ice Shelf on February 23, 2002, during large-scale collapse. Courtesy NASA/Goddard Space Flight Center Scientific Visualization Studio.

colossal shards of white glass swept together (figure 1.2). But it was just the beginning.

As the days progressed, more meltwater ponded and flowed into the fractures on what remained of the Larsen B Ice Shelf. Crevices became deeper and wider. Then suddenly, all of the meltwater drained away. A tumultuous sound rang out and a huge portion of the ice shelf broke free; other parts simply disintegrated. It was as if a monstrous bite, the size of Rhode Island (more than 3,000 square kilometers), had been taken out of the Larsen B Ice Shelf. By March, the ice shelf was essentially a floating expanse of icebergs and slush. Scientists across the world were shocked.

The 2002 collapse of West Antarctica's Larsen B Ice Shelf was unprecedented in modern history. Researchers are still trying to piece together exactly how it happened. A smaller collapse had occurred in 1995. For scientists studying ice sheet dynamics, these events were a game-changer.

The loss of ice at the Larsen B Shelf did not directly affect sea level (only land-based ice or snow, melting and flowing into the ocean, adds to its volume). However, as would be shown by measurements years later, the 2002 collapse released the brake on the land-based glacier behind it, allowing the river of ice to flow six times faster toward the sea.

The scale and speed of the 2002 Larsen B Ice Shelf breakup were startling. But what it and the 1995 event suggested was even more worrying: similar processes could play out on other, larger ice shelves.

And they already are. Ice is also fracturing and melting at the even more massive shelves buttressing the Pine Island and Thwaites Glaciers. In fact, a giant cavity beneath the Thwaites Glacier was recently discovered. It is two-thirds the area of Manhattan, about 40 square kilometers, and 300 meters deep. Scientists estimate that billions of tons of ice were lost within just the previous three years, an indication that the glacier is melting even faster than previously thought. The complete melting of the Pine Island and Thwaites Glaciers could potentially raise the ocean by more than 3 meters. If West Antarctica's Ross Ice Shelf (about half a million square kilometers, or the size of Spain) were to collapse and release the glaciers behind it, the flow of ice to the ocean could raise sea level another 3 meters or more.

In July 2017, a Delaware-size chunk of the Larsen C Ice Shelf, some 6,500 square kilometers, broke away. Scientists took note because it happened in the Antarctic winter. It is unclear what will happen next at the Larsen and other ice shelves in Antarctica, but very close attention is being paid, especially to those fronting large land-based glaciers.

The Earth's climate is changing. Among the big unknowns: How much of the vast expanses of snow and ice in Antarctica and Greenland will melt? How fast, and by what processes? And how far and how fast will sea level rise because of it?

Climate Change: The Known

Weather and climate have affected humankind for as long as our species has inhabited the planet. For much of that time, myth and folklore were used to explain the vagaries of the atmosphere or to predict its behavior. Today, scientists have real-time access to weather stations across the globe, along with data from satellites, ships, custom-equipped aircraft, weather balloons, ocean-voyaging and stationary buoys, and remotely operated vehicles. As never before, we are observing and monitoring the Earth and its atmosphere. Yet critical questions remain. To better understand what remains unknown, it is helpful first to consider some aspects of the Earth's climate that are well understood.

The Atmosphere and Carbon Dioxide

The atmosphere. It extends from the planet's surface to the edge of space and is only about 100 kilometers thick. Compared to the Earth, whose radius is nearly 6,400 kilometers, the atmosphere is wafer thin (see plate 1). Yet it is this thin layer of nitrogen, oxygen, and trace gases that provides for our every breath and prevents the planet from plunging into a frigid Mars-like cold. But the atmosphere is not an immobile source of life—it changes over time and space. Its never-ending fluctuations give rise to the day-to-day changes that form our weather and the longer-term variations that constitute climate. And variations in the makeup of the atmosphere influence and can drive change in the Earth's climate. This is not a big unknown, like the nature of dark matter, whether life exists on other planets, or what Batman wears under his tight-fitting rubber suit. For hundreds of years, scientists have been studying the Earth's atmosphere and how it affects our planet.

In the early 1800s, mathematician and physicist Joseph Fourier recognized that as the sun's energy or radiation passes through the atmosphere and strikes the Earth's surface, it heats up the planet. Without the atmosphere, though, the planet would regularly turn frigid. Fourier was the first to recognize that the atmosphere insulates Earth from heat loss—like a blanket. Then in 1859, scientist John Tyndall discovered something astonishing about one of the trace gases in our atmosphere—carbon dioxide. While the other major components of the atmosphere, nitrogen and oxygen, are essentially transparent to long-wave radiation, carbon dioxide is not. Carbon dioxide, along with water vapor, even in small quantities, absorbs long-wave energy, which is stored as heat. Several decades later, Swedish chemist Svante Arrhenius went further, suggesting that increased levels of carbon dioxide in the atmosphere could alter Earth's surface temperatures. Since that time, observations and experimental evidence have repeatedly confirmed these early discoveries.

Here's how it works. Incoming solar radiation (short-wave) passes through the atmosphere and strikes the Earth. Some of this energy is reflected back, especially from light-colored surfaces like ice or snow. But much is absorbed as heat and then re-emitted as longer-wave infrared radiation (we don't see such energy, much like ultraviolet light). Somewhat like the glass in a greenhouse, carbon dioxide, water vapor, methane, and other gases trap (absorb and re-emit) this long-wave energy as heat in the atmosphere. Again, some is lost to space, but much of the absorbed heat is directed back toward the planet—warming the air, ocean, and land.

The result of heat-absorbing greenhouse gases in our atmosphere: a fertile, warm Earth versus desolate, frigid Mars. But there's a catch. Humans are at times too smart for their own good. We discovered the power (pun intended) that comes from burning fossil fuels. And when fossil fuels are burned, they release additional carbon dioxide into the atmosphere, and more carbon dioxide captures more heat.

Ever wonder why they are called fossil fuels? Hundreds of millions of years ago on a very warm Earth, algae and other simple plantlike organisms flourished. After these carbon-based organisms died, some were buried deep beneath the land and seas. Over time, with decomposition, pressure, and heat, they transformed into oil, natural gas, and coal. These fuels are thus the preserved remains of prehistoric plants and other organisms—fossils. When fossil fuels are burned, the carbon they contain combines with oxygen, and carbon dioxide is released into the atmosphere—the very same gas that Tyndall and Arrhenius first showed traps radiant heat.

The burning of fossil fuels is not the only way humans add carbon dioxide to the atmosphere. But it is by far the largest anthropogenic source of carbon dioxide, followed by deforestation. Natural sources include the decomposition of organic material, volcanic emissions, weathering of rocks, respiration, and processes within the oceans.

Our best modern record of carbon dioxide concentrations in the atmosphere comes from the observatory at Mauna Loa in Hawaii. In the 1950s, Scripps Institution of Oceanography scientist Charles David Keeling and his colleagues began measuring the concentration of carbon dioxide in the atmosphere there and at other locations. Early on, they discovered a small daily variation in carbon dioxide concentrations. In the daytime, plants take up carbon dioxide through photosynthesis, then at night they release it via respiration. Later measurements revealed a similar seasonal variation. Carbon dioxide in the atmosphere decreases in the late spring and summer as it's taken up by growing plants. In the winter, carbon dioxide concentrations increase in the atmosphere because some plants die and decompose, and the release of carbon dioxide through respiration is greater than photosynthetic uptake. Over time, another and more startling pattern in the Mauna Loa carbon dioxide data became apparent: since industrial times the amount of carbon dioxide in our atmosphere has been rising, from about 300 parts per million to more than 400 parts per million.

Concurrently, the Earth's average temperature has risen more than 1°C since 1880. More important, the pace of warming has accelerated since 1950, with the last several years being the warmest ever recorded. Today, whether you look at the atmosphere or the ocean, at direct measurement or satellite data, the same tale is being told: carbon dioxide in the atmosphere is increasing and the climate is warming at a rate unprecedented in modern times.

People often argue that Earth has, throughout its history, gone through cycles of cold and warmth. Why then is today different from the past?

Our record of instrument-measured temperatures goes back only about 100 to 150 years. To compare today's rate of warming or current concentrations of carbon dioxide with those of the more distant past, scientists must find indicators or proxies that record previous atmospheric conditions. These include plants or other organisms that are sensitive to temperature or other climate variables and preserve records of their growth over time, such as corals, trees, and foraminifera (small shelled marine organisms). Bubbles of air trapped within layers of ice, undisturbed layers of sediment in lakebeds or oceans, and accumulations of ice, dust, pollen, and volcanic ash may also been used to establish prehistoric temperatures, dates, and carbon dioxide concentrations.

By combining the data from such indicators and from modern observations, scientists are able to reconstruct a record of global temperatures and carbon dioxide concentrations going back hundreds of millions of years. The detail or resolution diminishes as you go further back in time, but even so the data reveal a great deal about Earth's distant past. For instance, based on data from Antarctic ice cores, over the last eight hundred thousand years and up until about the 1950s, carbon dioxide concentrations in the Earth's atmosphere has varied between about 170 and 300 parts per million.

But going way back, some fifty million years ago, data indicate the concentration of carbon dioxide in the atmosphere was about 1,000 parts per million. Back then, there were no ice sheets, temperatures were 8 to 12°C warmer than today, and sea level was some 75 meters higher—the Earth was definitely less hospitable than it is today. Some three million years ago, carbon dioxide concentrations were similar to what we see today (350 to 400 parts per million). Temperatures were 1 to 3°C warmer than now and sea level was up to 20 meters higher. Again, it was not a very hospitable world for modern society as we know it. During the last peak interglacial (warm) period, 125,000 years ago, carbon dioxide concentrations were about 300 parts per million, temperatures were slightly warmer (1 to 2°C), and sea level was some 5 meters higher than today. The takeaway: in Earth's past, when carbon dioxide levels in the atmosphere were higher than or at levels similar to today, the planet was warmer, more sea-covered, and certainly a less roomy, hospitable home for its residents.

So what drove climate change millions or even thousands of years ago? It wasn't anthropogenic releases of carbon dioxide. Data suggest that back then the climate system was driven by other factors, including changes in solar output, massive volcanic emissions, the distribution of land and sea, ocean circulation, and tectonic upheavals followed by weathering. On time scales of about ten thousand to a hundred thousand years, orbital variations play an important role, particularly in forcing glacial and interglacial periods; these include the varying shape of Earth's elliptical orbit about the sun, the tilt of the Earth's rotational axis, and the wobble of the Earth's spin. Carbon dioxide played a role as well, but rather than driving change it appears to have either enhanced or reduced the effects of the other variables through feedback mechanisms. Today, these same forcing factors are present, but they are being overshadowed by the influence of carbon dioxide in the atmosphere.

Since 1950 the concentration of carbon dioxide in the atmosphere has risen to greater than 400 parts per million—a level far higher than it has been in hundreds of thousands of years. To investigate in more detail how Earth's climate has responded to this increase, scientists have reconstructed a precise record of temperature change for the last thousand years using both global proxy data and observations (figure 1.3). Here the data is especially revealing. In the Northern Hemisphere, a long-term gradual decline in temperature continued until about a hundred years ago. Then something happened. The temperature began to increase and at an accelerated pace. There have been similar periods of warming in Earth's past, following ice ages, but the amount of warming that has occurred over the last century, back then took ten times longer, on average a thousand years. So here's the point: It is not the actual temperature that is the issue. Rather it is the

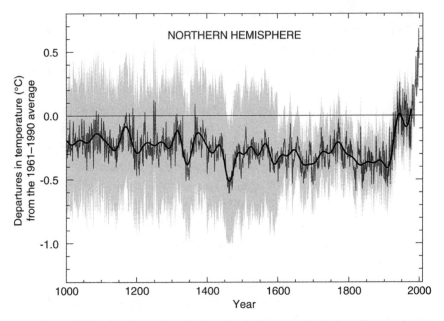

Figure 1.3. Temperature in the Northern Hemisphere over the last one thousand years based on proxies and, for roughly the last century, instrument observations. M. E. Mann et al., *Geophysical Research Letters* 26, no. 6 (1999), as shown in Summary for Policymakers, Third Assessment Report of the Intergovernmental Panel on Climate Change (2001).

rapid pace at which the global thermometer is rising that is unusual and problematic.

Another line of evidence indicating that the recent rise of carbon dioxide in the atmosphere is not natural comes from geochemists who study the isotopes of carbon. All atoms of a chemical element have the same number of protons; where isotopes differ is in the number of neutrons, and thus in atomic weight. Carbon has three naturally occurring isotopes, ^{12}C, ^{13}C, and radioactive ^{14}C. Scientists can measure and distinguish the amount of each in samples of carbon, carbon dioxide, and the atmosphere.

As plants grow, using carbon dioxide from the atmosphere in photosynthesis, they preferentially take up the lighter isotope of carbon (^{12}C) rather than the heavier (^{13}C). When plants are burned, more ^{12}C is then released into the atmosphere. Measurements since the Industrial Revolution show that the ratio of ^{13}C to ^{12}C in the atmosphere has decreased—the relative amount of "light carbon" has risen. Data on the amount of emissions and deforestation support the conclusion that most of the ^{12}C-rich carbon in the atmosphere comes from the burning of ancient plants, a.k.a. fossil fuels.

To further understand, and possibly forecast, future climate change, scientists also use computer simulations. People sometimes argue that computer models or simulations are unreliable and inexact. Yet computer modeling is used every day in a huge variety of ways: to control and improve traffic flow, to design and test safety standards, in construction, in healthcare, and in filmmaking. There'd be no Groot in *Guardians of the Galaxy* or Pandora in *Avatar* without sophisticated computer modeling and technology. In climate studies, modeling is based on the physical laws and processes that drive Earth's atmosphere, as well as oceanic processes. Simulations start with initial conditions derived from observed and/or proxy data and are run, tested, and calibrated. The results or outcomes are then compared with real-world conditions.

By modeling the forces that drive climate and including natural inputs of carbon dioxide to the atmosphere, scientists have been able to reliably simulate past conditions; the results, that is, closely replicate data from proxies and observations. When the same models are run for current conditions using only natural inputs of carbon dioxide, the accelerated pace of warming cannot be recreated. Only when the additional carbon dioxide released from the burning of fossil fuels is added do the models produce or recreate today's accelerated pace of global temperature rise. These same models have been run, extending into the future, with various levels of carbon dioxide in the atmosphere. Results suggest that with business as usual, by 2100 global temperatures will rise another 2°C or more.

Data clearly indicate that carbon dioxide in the atmosphere is increasing, mainly due to human influences, and that Earth's surface temperatures are rising unnaturally fast. Here then is where the big unknowns come in: at what rate will the climate continue to change, and just how extreme will the consequences be.

Climate Change: The Unknowns

MIAMI BEACH, 2016. The skies are a cloudless blue with the haze of high humidity. It hasn't rained in days. Strangely, though, there's water in the streets and it's rising. Water has pooled, several feet deep, in low-lying intersections, and it's flowing into underground parking garages, creeping up driveways, and beginning to bubble up through lawns and sewer drains. It's a nasty mix—seawater combined with the wastes and chemicals associated with a densely populated area. Residents and visitors alike shake their heads in disgust and, increasingly, in concern about their drinking water, which comes from an underground aquifer now at risk of contamination.

Since 2006 in Miami, flooding associated with rain events has increased 30 percent, and flooding due to tides has risen a whopping 400 percent. Here, the cause is no mystery: we know exactly why it is happening. On average, Miami sits about two meters above sea level, but much of the city and coast are lower. The bedrock underlying much of South Florida is limestone, or calcium carbonate, having formed in the past when sea level was higher. When exposed to slightly acidic freshwater, as in rain or water percolating through the ground, this type of rock is unstable or can dissolve. Hence, the ground in South Florida is not only low; it is also, simply put, full of holes. Add to that a rising sea, and the explanation for so-called sunny day flooding is obvious. When the tides are exceptionally high or coincide with onshore winds, seawater becomes an invading enemy, especially in low-lying areas like Miami Beach. The ocean flows across low spots along the shore and pushes into and through the drainage system as well as the underlying limestone. Summer squalls or rain events exacerbate the situation. With the rising sea, the region is also more vulnerable to storm surge—one of the most dangerous and costly impacts of hurricanes.

But the situation is even worse, because for some reason sea level is rising especially fast in southeast Florida. Research has shown that prior to 2006, sea level rise in southeast Florida was similar to the global average since 1993 of just over 3 millimeters per year. After 2006 the average rate of sea level rise in southeast Florida rose to about 9 millimeters per year. That doesn't seem like a lot, but over time it adds up, particularly during storms or high tides. It is unclear why the ocean in southeast Florida is rising more quickly than elsewhere. One hypothesis is that it may be due to slowing of the nearby Gulf Stream, which could result from increased freshwater input in the Arctic (more on this later in the chapter). We know why the ocean is rising, but why southeast Florida's rise is so fast remains an intriguing and troubling question.

In response to the growing problem of sunny day flooding in Miami, the local government is spending hundreds of millions of dollars to build infrastructure, including a pumping system, to keep water out. Unfortunately, it is little more than a Band-Aid for a worsening long-term problem. As our climate warms, in southeast Florida and across the world, just how fast and how far will the sea rise?

Sea Level Rise: How Fast, How Far?

In a warming world, global sea level increases due to the thermal expansion of seawater and meltwater input from land-based glaciers and ice sheets. On a local or regional scale, sea level rise is also strongly influenced by elevation changes, due, for instance, to subsidence, tectonic events, or the elastic rebound of the land when weight is removed by ice melting or weathering. In 2018 Representative Mo Brooks of Alabama suggested that erosion of the land, such as at the White Cliffs of Dover, is causing global sea level rise. There is absolutely no scientific evidence to support his claim; it was non-science and fake news—science style.

Across the globe, more than seven hundred million people live in low-lying coastal areas susceptible to flooding as a result of sea level rise. At greatest risk are people living in areas such as Bangladesh, the Louisiana delta, and the many small island nations.

During the twentieth century, data indicate that global sea level rose mostly due to the thermal expansion of the ocean and the melting of glaciers. Starting in about 1993, the pace of global sea level rise increased from about 2 millimeters per year to a little more than 3 millimeters per year. Research suggests that this acceleration is due to increased melting of land-based ice, especially in Greenland and Antarctica. In 2014 Representative Steve Stockman of Texas, referring to ice melting and sea level rise, noted in a hearing that "if your ice cube melts in your glass, it doesn't overflow. It's displacement." That's true, but we're not talking about floating ice (as in a drink). At issue, rather, are land-based glaciers and ice sheets melting and flowing into the sea. If you pour water into your glass from somewhere else, the level will rise, and potentially overflow.

Whether carbon dioxide emissions continue unabated, lower slightly, or are increased, according to extrapolations based on models and observational data, the sea is going to continue to rise. Predictions of how much vary widely—anywhere from 2 centimeters to more than 2 meters by 2100. But recent events and data have led many scientists to believe that most predictions underestimate the rate and extent to which the ocean will go up. One big thing we don't know and wish we did is just how much and how fast the world's great ice sheets and glaciers will melt and contribute to sea level rise.

The World's Large Ice Sheets

With improvements in technology and monitoring over the last two decades, experts are learning just how fast change in polar regions can occur. **—Ted Scambos**, glaciologist, National Snow and Ice Data Center

Large land-based masses of ice and snow exist today at both the southern and northern extremes of the planet, principally in the ice sheets and glaciers of Antarctica and Greenland. When summer loss of snow and ice exceeds winter accumulation in these regions, the net result is less frozen water. And when water that was originally frozen on land enters the sea, the ocean rises. Neither entire ice sheets nor all of the glaciers need to melt to drive sea levels dangerously higher. Even a meter of rise in the oceans would inundate major cities and coastlines across the world, cause economic devastation, and create a global refugee crisis.

Today, melting and net loss of ice is occurring in both Antarctica and Greenland. But the process of how large ice sheets, shelves, and glaciers melt is not well understood—no one was here when it last happened millions or hundreds of thousands of years ago. So teams of scientists from across the world are going to great lengths to monitor and understand what is happening.

Conducting research in polar climes is enormously challenging, which is one reason we know less than we'd like about the processes involved in the melting of ice sheets and glaciers. Brutally harsh weather conditions and geographic remoteness are just the tip of the iceberg. Research ships face blockades of sea ice, shifting icebergs, and ice-filled fjords, not to mention uncharted rocks or shoals. Add to that mechanical problems and logistical difficulties, which are the norm. Ice sheet research is also hampered by the simple fact that huge masses of ice and snow cover what scientists want to see and measure. Over the last several decades many of the challenges have been overcome through advances in technology.

Today, scientists studying the world's frozen masses have a wide range of tools at their fingertips. Aircraft and satellites provide amazingly detailed, high-resolution images of ice-covered continents and seas, and carry sensors that detect ice and ground elevations, variations in gravity, surface color or reflectivity, and surface temperatures. University of Colorado researcher Jim White notes that gravity measurements in particular are important when studying large ice sheets. Gravity allows scientists to determine the density of the ice, not just its extent or height—an important parameter in calculating ice loss over time.

To investigate inaccessible ice-bound oceans, specialized aircraft deploy dropsondes—instruments that, dropped from an airplane, collect atmospheric data as they descend. Drill rigs that can bore through hundreds of meters of ice allow the history of glaciers and subsurface features to be studied. Instruments linked to the global positioning system (GPS) can be left in place on or below the ice to track ice or ground movement. To view what lies hidden beneath and within the ice, scientists have ice-penetrating radar. Icebreaking ships outfitted with multibeam sonar and oceanographic sampling gear provide more detailed views of the ocean and seafloor, near and under the ice. And remotely operated vehicles or gliders are a new means of exploring hard-to-access under-the-ice environments, while aerial drones offer a new perspective from above. In terms of technology, it is an exciting time to be pursuing the secrets of the frozen but melting world.

ANTARCTICA. Covered 98 percent by ice, Antarctica hosts the coldest, windiest, and arguably the most remote and least hospitable land on Earth. Strong winds, up to 350 kilometers per hour, whip across the continent. In the winter, air temperatures can fall to a literally freeze-your-butt-off −98°C (as measured in small ice valleys in East Antarctica). Antarctica is encircled by strong ocean currents, high seas, and, in winter, sea ice. It is among the most extreme and difficult places to access and study. Yet to better understand the changes taking place in Antarctica, scientists and explorers have long fought against the elements. And while we now know more about Antarctica than ever before, giant unknowns remain.

Antarctica's ice sheet is the largest on Earth, some six and a half times larger than that of Greenland and, at its thickest, as deep as the Alps are high—nearly 5 kilometers. The average elevation in Antarctica is 2,300 meters above sea level, but in some places the land beneath the ice is equally as far below sea level. Antarctica is typically divided into West and East Antarctica (separated by the Transantarctic Mountains) and the Antarctic Peninsula. An extensive network of ice streams and glaciers ring the continent and bring ice from the interior to the coast (figure 1.4). Where glaciers flow beyond their underlying attachment to the land, floating shelves of ice form. Antarctica has fifteen major ice shelves, each tens of thousands of kilometers across and hundreds of meters thick. In the winter, thinner sea ice, just 1 to 2 meters thick, forms and extends seaward but melts come summer.

Geological data, going back millions of years, indicate that when the planet is at its warmest, Antarctica is ice-free and global

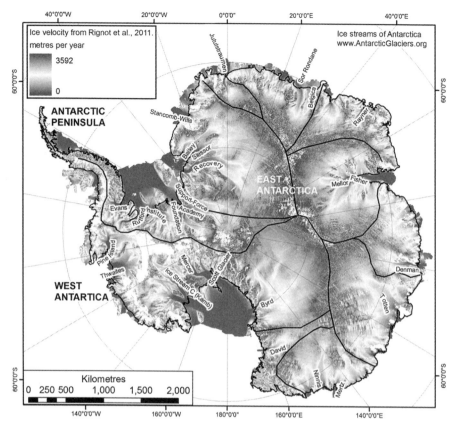

Figure 1.4. Ice streams and ice velocity in Antarctica. Courtesy Bethan Davies, www.AntarcticGlaciers.org.

sea level can be more than 60 meters higher than today. A wealth of data now shows that Antarctica is losing ice and that melting has accelerated since 2008. But the pattern and processes of change across the vast, ice-bound continent are complex and vary from place to place. While snow appears to be accumulating on East Antarctica's ice sheet, many glaciers in the same area are thinning, shrinking, and retreating inland. Glacial retreat occurs when the terminal end of a glacier moves inland due to melting, calving, or collapse.

In West Antarctica, the data are very clear: air and ocean temperatures are rising, and glaciers and ice shelves are melting, thinning, retreating, and collapsing. On the whole, ice loss in West

Antarctica more than makes up for increased snow accumulation in the East. If all of West Antarctica's ice were to melt, it would raise sea level some 5 to 7 meters. But again, even 1 meter of sea level rise is enough to cause catastrophic changes to the global landscape.

Ice is warmed; it melts. Seems pretty simple. Not exactly, when we're talking massive ice sheets and flowing glaciers on the land, and thick ice shelves that float over a moving, changing ocean. Atmospheric dynamics and the configuration of the underlying bedrock add to the complexity. Presented here are a few discoveries about the melting of Antarctica's vast plains, shelves, and rivers of ice to highlight some of the big questions being asked, the complexity of the processes involved, and the worrisome nature of what has been learned so far. The foundation for these studies lies in decades of previous achievements in exploration, science, and technological development.

Unfortunately, there is, at present, no long-term record of how meltwater forms and drains off the Antarctic ice. However, in 2017 scientists released the results of a continent-wide survey of summer surface drainage patterns based on satellite and aerial imagery taken from 1947 to 2015. Hundreds of meltwater ponds and streams were identified and mapped. One of the most visually stunning discoveries was made on the southern Nansen Ice Shelf. Here, the flow of meltwater ended in a towering waterfall 120 meters wide (plate 2). On the Amery Ice Shelf in East Antarctica, a network of streams 120 kilometers long was found connected to a series of ponds up to 80 kilometers in length. The extent of melting and the drainage network surprised even the scientists, occurring farther inland and at higher elevations than expected. Some of the features discovered had been in place for decades, while others appear more recent in origin.

Many of the meltwater streams mapped in the study were found on slow-moving glaciers adjacent to areas of exposed dark rock or older, air-free "blue ice." Darker surfaces absorb more of the sun's radiation than those that are lighter (and thus more reflective), and this can enhance melting. Wind scour or glacial thinning can expose the underlying dark rock or blue ice, and thus foster melting.

Not surprisingly, the slope of the ice shelves was found to play a role in drainage. More numerous and distinct ponds and streams, which may promote fracturing, were found on relatively flat ice shelves. On steeper slopes, ponds were less common; instead, networks of rivers appear more likely to form and transport meltwater to the ocean more quickly. Overall, the study revealed that extensive, long-standing surface meltwater ponds and streams are much more widespread on Antarctica's ice than was previously known.

As often happens in science, this research highlighted lingering unknowns, while also raising new questions. Will meltwater ponds and streams in flat areas contribute to ice shelf fracturing and instability, as appears to have been the case in the Larsen B Ice Shelf collapse? Do long-standing streams on steeper slopes more efficiently transport meltwater to the ocean and help to avoid weakening of the ice shelves? Are there long-term trends in the processes and amount of melting and drainage?

In 2016 especially widespread surface melting was detected in the Ross Sea area of West Antarctica. It may have been one of the largest melt events in the region since 1978. Scientists suspect the cause was warmth combined with a strong El Niño that brought rain and warm air to the region. (For more on El Niños, see chapter 4.)

At Pine Island Glacier in West Antarctica, scientists made another startling discovery. The first surprise came when they found warmed meltwater flowing from beneath the glacier's base. When they got a look at the underside of the glacier's ice shelf, something even more astounding became apparent. The meltwater flowing out from below the glacier was warm enough to carve canyons in the underside of the shelf. Pine Island Glacier was not only melting from above but also being scoured out from below. And that's not all that's going on below the Pine Island Glacier. Data indicate that relatively warm ocean water is also lapping up under the thick and floating ice shelf. Here, deep down, the ocean isn't exactly toasty, but with the effects of pressure on ice, which effectively lowers its melting point, seawater temperatures are warm enough to cause melting.

Surveys around Antarctica have documented that over the past forty years, though slightly modified by mixing processes, the deep ocean is indeed warming and increasingly flowing into shallower depths near the coast. Submarine melting is causing rapid ice loss at the Pine Island Ice Shelf and throughout West Antarctica's ring of ice shelves.

If the relatively warm ocean water penetrates far enough inland, it could cause a new problem. Much of the bedrock in West Antarctica lies below sea level, and if warm water spills into these areas, under the ice, some scientists argue it will trigger unstoppable melting. Some say it is already happening. Changes in wind patterns around West Antarctica could bolster this process, as stronger offshore winds can intensify circulation and cause more deep water to well upward. Research suggests that this too is already occurring.

Compared to West Antarctica, East Antarctica's much larger ice sheet, glaciers, and floating ice shelves have long thought to be more stable. Is this still the case?

The Totten Glacier is the largest feeder of ice to the ocean from East Antarctica's immense ice sheet. It is some 65 kilometers long and 30 kilometers wide, and holds enough ice to raise global sea level by more than 3 meters if melted. Based on satellite imagery, scientists have discovered that between 1996 and 2013 Totten Glacier's elevation dropped more than 12 meters and its seaward margin retreated inland some 3 kilometers. Totten Glacier was thinning faster than any other ice flow in East Antarctica. It was a shocking result and, at the time, a mystery. What was the cause of such rapid ice loss?

In 2015 an opening in the sea ice allowed, for the first time, an Australian ship to approach the edge of the Totten Glacier Ice Shelf. Before the ice could refreeze and trap their vessel, researchers quickly investigated the configuration of the ice shelf, the underlying bedrock, and the adjacent ocean waters. Their results revealed that beneath the Totten Glacier Ice Shelf lies a 10-kilometer-wide submarine canyon with depths approaching 1,100 meters. Landward, the canyon branches into two even deeper, but narrower, channels. Using data from oceanographic profiling instruments, the scientists found that these deep, narrow channels act as chutes or funnels for relatively warm deep ocean water.

The cause of Totten Glacier's unusually rapid flow, retreat, and thinning was now apparent—submarine melting. It is not just a West Antarctica phenomenon but is happening in East Antarctica as well. This suggests that other ice shelves and glaciers in the region may also be melting at their bases due to warm ocean water. And here too, the strengthening of offshore winds appears to be causing more warm deep water to well upward. As in West Antarctica, a large fraction of East Antarctica's bedrock is below sea level, thus creating the potential for similar runaway and unstoppable melting.

A new twist in the Antarctic ice melt puzzle has arisen—literally. The removal of ice and its weight from the land can cause the ground to slowly rebound upward. Recent research indicates that as the ice melts in Antarctica, the ground is rising, and more quickly than expected. Data also suggest that the Earth's underlying mantle may, in this area, be more fluid than elsewhere, which would explain the faster-than-expected rebound. The rising land could help to stabilize the ice above and prevent catastrophic collapse—maybe.

Along with melting and the rising land, there's another intriguing under-the-ice occurrence in Antarctica—lakes. Some four hundred lakes are known to exist beneath Antarctica's ice. Lake Vostok, the largest and most famous subglacial body of water in Antarctica, sits beneath nearly 4 kilometers of ice and is 500 meters deep and 250 kilometers long. Background geothermal heat and the pressure of the overlying ice allows the lake to remain liquid even though its temperature is believed to average a frigid −2°C. Several attempts to drill through the ice down to Lake Vostok have failed due to accidents and contamination. In 2015 a Russian team claimed to have successfully sampled Vostok's water, looking for evidence of extreme microbial life. So far, reports are sketchy. Scientists want to know more about the lake water buried beneath Antarctica's ice: Does it host alien-like microbes capable of living in one of the darkest, coldest, most extreme environments on the planet? And how do subglacial lakes influence ice dynamics, such as drainage and elevation?

Scientists have learned a lot about the fracturing and collapse of ice shelves since the Larsen B event in 2002, but questions linger. What is the exact role of hydrofracturing—where infiltrating meltwater creates or widens crevasses? How fast does it usually (or unusually) happen? Another intriguing and worrying question is about the stability of giant ice cliffs. When a major calving event or collapse occurs at an ice shelf, it removes the compressive forces provided by the massively thick ice—a bit like the arches of a bridge. Are newly exposed ice cliffs, particularly those more than 90 meters in height, inherently unstable, and could they collapse under their own weight? Will this exacerbate the speed by which ice shelves and glaciers fall apart and retreat? It is a big wish-we-knew and researchers are trying to find out.

Around Antarctica, there's another mystery frozen in the ice—this time in relatively thin ice. Each winter, sea ice, about 1 to 2 meters thick, forms at the edge of the continent's ice shelves and extends seaward. While overall, Antarctica's ice is melting, over the last decade or so its winter sea ice has been expanding, especially in the southern Ross Sea area. Scientists theorize that the expansion may be due to the strengthening of cold winds blowing across the surface or from ice shelf melting. Cold freshwater melted from the ice shelves lowers the salinity and temperature of surface water, making it easier to freeze. However, the story is complicated. In 2016, 2017, and 2018 the maximum extent of Antarctica's winter sea ice decreased. Scientists are waiting to see what will happen in the coming years.

The ice in Antarctica is melting; that is not an unknown. It is happening faster than once thought possible, through processes that remain difficult to study and are not fully understood. As climate change continues, scientists will be probing, monitoring, and studying Antarctica's vast ice sheets, glaciers, and frozen shelves. Their data will help us to better understand how much ice is being lost and at what speed—which will, in turn, help us predict how fast and how far sea level will rise. But Antarctica isn't the only place where the melting of ice is contributing to a rising global ocean. Currently, it is estimated that about a third, or 1 millimeter per year, of global sea level rise, is coming from the melting of the second largest ice mass on the planet.

The Arctic

ILULISSAT, GREENLAND 2017. A deep whumpf echoed across the Kangia Icefjord. Somewhere among the fjord's great sea of icebergs or along the glacier's massive floating shelf, a gargantuan chunk of ice had given way. It's a regular event in this, the outlet of one of the world's fastest flowing rivers of ice—the Semaq Kujalleq (Jakobshavns Isbræ in Danish) glacier (figure 1.5). In the summer of 2012, the ice here was clocked moving at a startling 46 meters per day. Typically, the Semaq Kujalleq glacier flows at a still-quick 20 meters per day and drains more than 6 percent of Greenland's 1.7 million-square-kilometer ice sheet. The glacier and its floating ice shelf, hundreds of meters tall, are also among

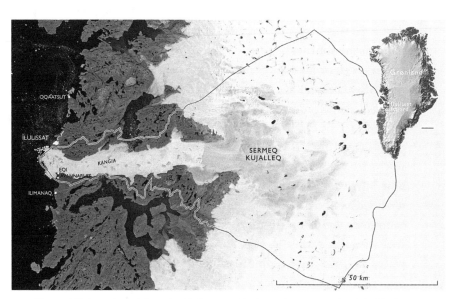

Figure 1.5. Overview of Greenland's Sermeq Glacier and Kangia Icefjord. © GEUS, Geological Survey of Denmark and Greenland/Landsat 7.

the planet's most prolific producers of icebergs. And nowhere is this wonder better observed than after a short hike from the small village of Ilulissat.

Ilulissat sits nestled amid ancient fjords midway up the west coast of Greenland. Its residents have long relied on the nearby glacier and the immense productivity it brings to the surrounding sea. During the winter months, the land is frozen solid and sea ice forms within the Kangia Icefjord. With summer's warmth comes melting and the release of nutrients into the ocean. These nutrients trigger blooms of plankton that create a bountiful food web supporting fish, seals, and whales. But such biological wealth is not the only thing in abundance in summer; so are icebergs. It is the place to go if you want to see calving ice, crevassed ice, blue ice, striped ice, and more. The town is increasingly a destination for so-called clima-tourists—people who want to see climate change up close and personal.

Warmth transforms the Kangia Icefjord into a shifting icy marvel (plate 3). Lined by dark prehistoric rocks, the narrow gorge becomes choked with ice—an extravaganza of icebergs, bergy bits (less than 5 meters in size), growlers (less than 2 meters in size), and innumerable smaller pieces. The icebergs range from mountainous peaks to boxy tabular forms, angular, domed, or gently sloped like a beginner's ski slope. They can sit a hundred meters high or lay flat, barely breaking the sea surface. As melting and calving eat away at the ice, strange shapes take form: tall, skinny ice mushrooms; turreted spires; broad arches (figure 1.6). The surface of the ice also varies; it may be smooth to wavy or made jagged by side-by-side crevasses. Glacial sediment entrained in the ice creates brown stripes in many of the fjord's icebergs. This "dirty" ice absorbs the sun's radiation better than the white, air-filled ice and melts more quickly, creating deepening fractures. Summer's warmth also creates crystal clear pools of turquoise meltwater in surface depressions on the icebergs. Where meltwater has refrozen or compression has squeezed the air from the ice, translucent blue hues are found in patches, layers, or stripes. It's not just the summer warmth or fast-moving glacier that creates this astonishing sight; part of the cause lies downstream.

Figure 1.6. Iceberg shed from Sermaq Glacier and Kangia Icefjord. Photo E. Prager.

At the Kangia Icefjord's mouth, where it empties into Disko Bay, there's a major bump in the road for icebergs traveling to the open sea. It is a rocky moraine, or ledge, bulldozed and left in place by the glacier thousands of years ago. Here, the depth abruptly shallows from 1,000 meters below sea level to less than 200 meters. Driven downstream by wind and currents, icebergs that extend deeper than 200 meters become stranded. They in turn block the ice behind, creating a frozen but shifting traffic jam in the fjord. Smaller bergs and bits escape the ice-choked passage on almost a daily basis but, it can take days, weeks, or even years for a large iceberg to lose enough ice to float off and into Disko Bay.

Once free, the icebergs are pushed and pulled by the wind, outgoing tide, and ocean currents, often passing right by Ilulissat. They drift north driven by the West Greenland Current. The largest icebergs eventually head west toward Canada and then south in the Baffin Current. They can last for years and have been sighted at latitudes as far south as New Jersey. The Kangia Icefjord is even believed to have birthed the iceberg that sank the *Titanic* in 1912.

In 2004 the Kangia Icefjord was named a United Nations World Heritage Site. As a protected landmark, its stark beauty, historical significance, and scientific importance will be preserved. A new visitor and research center is being built to showcase both the site and a topic of more than local importance—climate change.

Since 2000, scientists have determined that 30 percent of the 739 gigatons of ice lost from Greenland has come from the Semaq Kujalleq glacier. "The glacier's calving events are becoming more spectacular with time," notes Eric Rignot, a glaciologist at NASA's Jet Propulsion Laboratory, "and the speed and retreat rate of the glacier are amazing." Since the 1990s the speed of the glacier's flow has doubled, while its terminal end has retreated dramatically (figure 1.7). Between 1850 and 2010, due to ice loss, the glacier moved inland 40 kilometers—in just the last ten years of that time, the retreat was more than in the previous hundred years.

In 2008, while filming for the award-winning documentary *Chasing Ice*, filmmakers recorded the largest calving event ever witnessed—on Semaq Kujalleq's ice shelf. For seventy-five minutes, blocks of ice hundreds of meters tall wrenched free of the shelf, overturned, and cascaded into the sea. In this one ginormous calving event, the five-kilometer-wide shelf is estimated to have retreated inland more than a kilometer.

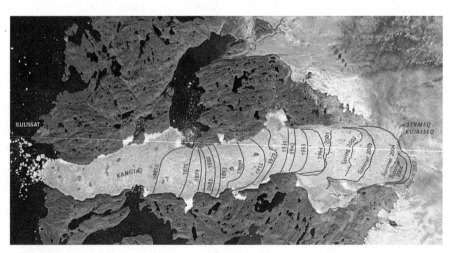

Figure 1.7. Retreat of Sermeq Glacier, 1850–2007. © GEUS, Geological Survey of Denmark and Greenland/Landsat 7.

Researchers studying the Semaq Kujalleq glacier (and the downstream Kangia Icefjord) have learned much about its great retreat and ice loss. The causes their data reveal echo those found in the Antarctic: surface melt, lubrication by meltwater at the glacier's base, submarine melting exacerbated by relatively warm ocean water flowing up and under the ice shelf, and calving and weakening of the shelf, which reduces its ability to act as a buttress. And as in Antarctica, the big unknowns here are what is going to happen next and how fast is it going to occur.

Some scientists predict that the Semaq Kujalleq glacier will produce even more icebergs in the future and may retreat some 50 kilometers by 2100. What would this mean for sea level rise? Will more icebergs threaten ship traffic, which is slated to increase as more ice melts and calves in the Arctic? What will it mean for the residents of Ilulissat and the ecosystems of the region? Researchers are working hard to find out.

Semaq Kujalleq is just one of the two hundred glaciers that move ice from Greenland's massive interior ice sheet toward the coast. Some of these glaciers reach the shore; others do not. Some have large ice shelves; others do not. The bedrock underlying the glaciers and ice shelves is also diverse, laced with a variety of sills, ridges, valleys, and channels. It is not one-size-fits-all for Greenland's glaciers. And as in Antarctica, the processes of melting are complex. But if all of Greenland's ice were to melt, it could raise sea level more than 7 meters.

Research indicates that in Greenland the air as well as the surrounding ocean are warming. Air temperatures have risen some 1.2°C since the 1890s. Scientists studying how Greenland's glaciers are responding have confirmed some long-held ideas about how massive areas of ice melt, but there have also been surprises. Here are a few more examples of what has been found, and some of the big unknowns that continue to drive the research community in and around Greenland.

Petermann Glacier in northwest Greenland has one of the Northern Hemisphere's oldest and largest floating ice shelves, hundreds of meters thick. In 2010 and 2012, giant islands of ice broke off the glacier's ice shelf, one of which was twice the size of

Manhattan. Together, the two calving events caused the glacier to retreat 35 kilometers inland. Research expeditions, on the glacier and aboard icebreaking ships, have since begun to unveil Petermann's icy secrets. Scientists have now mapped the bathymetry or depths of the underlying seafloor and sampled the ocean in and around the glacier's ice shelf. And in 2015 researchers installed an underwater instrument observatory. Three oceanographic stations were put in place, one under ice nearly as thick as the Empire State Building is tall. Data were recorded and sent automatically via a cable through the ice to a surface weather station and then via satellite to researchers.

The data from under Petermann's floating ice shelf is helping to reveal how the ice and ocean interact. For example, every two weeks, tides wash surface waters up to and away from the ice. These waters are cold and less salty due to ice melt. But beneath the surface, some 200 meters down, lies warmer, saltier water that originates in the North Atlantic Ocean. As elsewhere, this deep, relatively warm water was found moving rapidly onto the continental shelf—and under the Petermann Ice Shelf—where it is causing submarine melting and fostering rapid ice loss.

High-resolution satellite imagery has also revealed a large fracture on Petermann Glacier's ice shelf. Scientists have confirmed its location and extent using airborne imagery as part of NASA's long-term polar study, Operation Icebridge. The surprise here is that the large crack appears to have started on the interior of the ice shelf, not at the sides as expected. The significance of the crack remains the subject of study: Will it interact with other growing rifts and cause another large ice island to break off—suggesting that the shelf is on the verge of a large collapse? Or might it become stabilized over time? Using satellite imaging, scientists are keeping a close, but distant eye on the Petermann Ice Shelf.

At Petermann and Semaq Kujalleq glaciers, the ice shelves are melting and being weakened from below. To determine how prevalent the phenomenon of submarine melting is around Greenland, NASA initiated an intensive multiyear program named Oceans Melting Greenland (cleverly, OMG for short). So far, the results suggest that many of Greenland's glaciers terminate in deep water and are therefore vulnerable to submarine melting. Oceanographic sampling has documented that in the currents that circulate clockwise around Greenland, cold, less salty surface waters are mixing with warmer subsurface water, particularly in Northwest Greenland. And at the base of the glaciers, discharge of meltwater may entrain deeper warmer water and exacerbate submarine melting. As more data from OMG flows in, more will be revealed about how Greenland and the surrounding ocean are responding to climate change.

Another multiyear study is investigating how particles such as soot, dust, ash, and algae darken the region's glaciers and impact melting. Data from the Black and Bloom project, which began in 2017, are already revealing that algae and soot-covered surfaces enhance melting by reducing reflectivity and increasing radiation absorption.

While studying Rink Glacier on the west coast of Greenland, researchers got another surprise. It came in the data from a GPS sensor deployed on the bedrock beside the glacier—one of fifty such sensors positioned around Greenland to monitor changes on glaciers through ground deformation. Results revealed that in especially warm summers, the ice doesn't just thin or melt faster, it can actually slide as an enormous wave toward the sea.

Rink Glacier appears to flow on average about 2 kilometers a year. But during a warm four-month period in 2012, a wave of ice raced down the glacier at an average of about 7 kilometers per month. Over a gigaton of ice was put in motion. Scientists are certain that surface melting of snow and ice triggered the speedy flow, but the processes involved are unclear. Intense melting may have created ponds and rivers that eventually drained down through the ice and lubricated the glacier at its base. This, combined with melting and drainage that softened the glacier at its sides, could have enhanced its ability to flow. Researchers elsewhere are looking at similar all-out glacial surges, especially in mountain glaciers where rapid movement can pose a deadly threat to nearby residents.

Other studies in Greenland have noted that changes in the jet stream and high pressure over the Arctic are causing additional melting and that the summer melt season is starting earlier and lasting longer (more on changes in the Arctic later in the chapter).

Conditions in Greenland can be treacherous, research is difficult, and many wish-we-knews remain. For instance, what is happening in winter? But one thing is very clear: as the climate warms, Greenland is losing ice.

The ice in both Antarctica and Greenland is melting. It is being lost from the tops of glaciers as well as from below, at the sides, in giant fractures, and through colossal collapses. While climate change is the driving force, the color of the ice, strength of the winds, rivers of meltwater, upwelling of deep warming ocean water, and removal of buttressing cliffs and floating ice shelves all play a role. Though we still don't fully understand all of the processes by which massive ice sheets, shelves, and glaciers melt, the evidence suggests it is happening at an accelerated rate, at times in unexpected ways, and faster than previously predicted. What does this mean for the predicted rates of sea level rise? Stay tuned as more data and meltwater flow in.

SARICHEF ISLAND, ALASKA. The summers were short and cold, the winters long and brutal; and year-round the ground remained frozen. Thick sea ice protected the shore from storm waves and erosion. In the frozen north, life was difficult, but for at least four hundred years the Inupiaq people made the small island of Sarichef their home. Hundreds of people built houses and lived off the land and sea. They hunted, traveled, and resided on the frozen ground and ice. Then things began to change. Summers grew longer and warmer. Sea ice formed later and melted earlier, leaving the coast vulnerable to summer storms. Waves began to eat away at the coast, and homes began to fall into the sea (figure 1.8). And the frozen ground began to thaw.

On Sarichef, climate change is an in-your-face reality. In 2016 a vote was taken. The residents of Shismaref village elected to leave the island and relocate to the mainland. The frozen ground, or permafrost, their village was built on was and is disappearing. Thawing and storms now claim, on average, about 7 meters of shoreline annually. And with thinning ice, hunting has become

Figure 1.8. House collapse in Shishmaref, Alaska, due principally to coastal erosion along with issues of permafrost thaw. Courtesy *Nome Nugget*/Diana Haecker.

a dangerous endeavor. At least one man was killed when he fell through the ice on his way back from a hunt. Where he was lost, residents say, the ice should still have been solid. Some people are against moving the village to the mainland, worried it will change their traditional way of life. But the land is giving way and time is running out.

Shismaref isn't the only Alaskan village in trouble; some thirty-one others are also at risk as a result of climate change. But even when residents want to relocate, the costs can be daunting, if not prohibitive. In Shismaref, as a temporary fix, they've moved buildings inland and created rock and sandbag barriers at the coast. Eventually, as temperatures warm, the sea rises, and the ground continues to thaw and give way, residents will have no choice but to leave.

The Arctic typically refers to the region surrounding the North Pole and above the so-called Arctic Circle, an imaginary line wrapping around the globe at 66°34' north latitude; this includes ice-covered Greenland as well as parts of Russia, Canada, Scandinavia, and Alaska. All across the region, climate change is taking hold, and here it is happening faster than anywhere else on the planet.

Over the last fifty years, temperatures in the Arctic have increased twice as fast as the global average. From 1998 to 2012, warming in the polar north was as much as six times the global average. And from 1975 to 2012, the thickness of sea ice in the Arctic Ocean declined by some 65 percent. Most of the ice that forms now is seasonal, freezing in the winter and melting in the summer. With less sea ice, more radiation is absorbed by the ocean, enhancing or amplifying warming. And it's not just the loss of sea ice that is causing the Arctic's thermometer to rise. Heat and moisture transferred poleward, cloudiness, and shifts in the jet stream and storms are also thought to contribute to the planet's northerly warm-up. In 2018, Arctic temperatures were 17°C above average for days on end. Sea ice diminished rapidly. Scientists now predict that by 2030 (if not sooner) summers in the Arctic Ocean will be ice-free.

With so much of the old ice melting, the quality of the remaining ice is also a worry. In places like Alaska, the thickness of the ice, its stability, and its mobility impact ecosystems and wildlife as well as the lives and livelihood of people in the region. Where more mobile ice is pushed ashore, it creates impassable ridges and people can't collect driftwood for fires and heating. Thinner, broken-up ice, like slush, may look solid on a map or from above but be dangerously unstable and create hazards at sea. Ice dams can cause flooding. The amount and quality of ice impact industries as well, such as fisheries and transportation. In Alaska, salmon is king, the lifeblood of the state. Researchers don't yet know what impact climate change and ice loss will have on salmon. In fact, they don't even know where salmon go after entering the sea from the rivers of their birth and before returning to spawn. The unknowns here are abundant and have serious implications for people, wildlife, and the world.

Alaska's glaciers are also losing ice. Researchers with NASA's Operation Icebridge report that the state's glaciers are losing some 75 gigatons of ice each year and now contribute significantly to sea level rise. Numerous international efforts are under way to better understand and model the complex interactions between the atmosphere, ocean, land, and ice in the Arctic. Known as the world's refrigerator, the Arctic plays an important role in the distribution of heat on the planet, particularly for the Northern Hemisphere. Today, that refrigerator is on the fritz.

Permafrost

With regard to climate change, two of the big wish-we-knews in the Arctic have to do with permafrost: how much will thaw and how fast? Permafrost is defined as soil, rock, or sediment that has been frozen for at least two years. Some permafrost, however, is much older—the chilled remnants of the Earth's more ancient frozen past. It may be relatively thin, only about a meter in thickness, or extend more than a thousand meters deep, as in Siberia. Permafrost can be topped by a seasonally thawing layer or blanketed by tundra, forests, a lake, or the ocean. With more land than ocean in the northern extremes, most permafrost occurs in the Northern Hemisphere—some 23 million square kilometers of it. Extensive areas of frozen ground are found in, but not limited to, Canada, Russia, Alaska, China, and Greenland. Why is the thawing of permafrost so important and concerning?

Permafrost contains organic matter, the remains of dead plants and animals. When frozen, this organic material is well preserved, like the some forty-thousand-year-old wooly mammoth discovered in Siberia in 2013. When permafrost thaws the organic matter begins to decay (a very smelly wooly mammoth). As part of the decay process, microbes eat the organic matter and release either carbon dioxide or methane. And when it comes to warming, methane is a powerful greenhouse gas—like carbon dioxide on steroids, although not as long lasting. So as permafrost thaws, one question is which will be produced—carbon dioxide or methane? The answer is partly known. Where oxygen is unavailable, such as in anaerobic mud, wetlands, or swamps, microbes produce methane in the decay process. Less clear is where, when, and how much of the frozen ground will thaw and how much methane and carbon dioxide will be released.

A big part of the worry is that there's a whole lot of carbon frozen down there. Scientists estimate that the amount of organic matter frozen in permafrost represents more total carbon than is in the atmosphere today. Already there is ample evidence of permafrost thaw—take Shishmaref, Alaska, for example. And it is happening in many other places as well. In Canada, giant slabs of permafrost are slumping, causing landslides, and releasing large quantities of mud and silt into rivers and lakes. Similar collapses are occurring in Alaska, Siberia, and Scandinavia. In Alaska, as permafrost thaws, roads are buckling and trees tilting or falling over in what are called "drunken forests." In Siberia, a huge crater—a kilometer wide by 86 meters deep (and growing)—has opened in the permafrost. Scientists studying the crater believe the collapse was facilitated by deforestation in the 1960s. The exposed walls of the crater may offer a new geological window into climate going back two hundred thousand years.

Other evidence of permafrost thaw has risen to the surface—literally. Off Norway's Svalbard Islands, where warm ocean currents flow over the once-frozen seafloor, plumes of methane bubbles now rise toward the surface. In Siberia, methane is reportedly bubbling up through the ground and inflating thousands of foot-squishy mounds. Permafrost thaw poses a threat to the stability of the overlying ground whether it is a mountain slope, rolling plain, forested land—or an area built up with human infrastructure. In Alaska and Canada, roads are particularly vulnerable as the permafrost thaws. It also puts habitats and wildlife at risk.

But just mapping the extent of permafrost is difficult. Satellite sensors, for instance, typically look only at surface characteristics. So how do we monitor the extent or rate of permafrost thaw if we don't have a good map to start with or an efficient way to assess change over time and space? If only it was as simple as knowing the surface temperature. But thawing of permafrost depends on more than temperature, including how much ice exists within the frozen ground, the hydrology and topography of a given area, the ground composition and depth, and what's covering it. Vegetation, snow, or ice can insulate the permafrost below.

Not all of the methane or carbon dioxide released from thawing permafrost will make it to the atmosphere. For instance, methane-eating microbes in the ocean may take up much of the gas released from permafrost under the sea. But the potential release of methane and carbon dioxide is so huge—all that carbon below—that even with mixing or some removal, it could still greatly worsen warming. Another looming question: Is there a tipping point, when the air, ground, and sea are sufficiently warm to thaw enough permafrost that huge releases of methane or carbon dioxide are imminent and unstoppable? Wish we knew.

Some scientists propose that gas released from thawing permafrost will be offset by longer growing seasons in the Arctic and elsewhere, such that plants will remove the excess carbon dioxide or marine algae suck it up. But plant and algae growth are influenced by many other factors as well, such as access to nutrients and water, along with temperature. Recent research suggests that drought or insufficient soil moisture may limit carbon dioxide uptake by plants in a warming world.

The thawing of permafrost in the Arctic is already changing the landscape in ways never before observed. Scientists will continue to monitor, investigate, and wait to see how soon and how much permafrost will thaw. And permafrost thaw is not yet included in the computer models used to predict warming and sea level rise. It's another reason why the Earth may warm and the ocean may go up faster than predicted.

The Oceans' Big Flow

Another wish-we-knew regarding both the Arctic and climate change has to do with what is often referred to as the Great Ocean Conveyor Belt (plate 4). This is the global density-driven circulation of the oceans caused by changes in salinity and temperature. The system works something like this: In the Arctic, principally in the regions of the Labrador and Greenland–Norwegian Seas, the salinity of already-chilled seawater is increased due to evaporation by intense winds and the concentration of salt during ice formation. This cold, now saltier seawater is more dense and sinks into the deep ocean (below 1,000 meters). From there, it flows south and spreads through the North Atlantic Ocean. North Atlantic Deep Water, as it is called, then moves into the South Atlantic, around Africa, and eventually into the Indian and Pacific Oceans. An input of deep water from the Antarctic (Antarctic Bottom Water) adds to the cold, salty water traveling the oceans at depth. Over time and with mixing, this deep ocean water eventually warms, becomes less salty, and rises. The now less-dense seawater flows near or on the surface from the Pacific and Indian Oceans back into the South Atlantic. It then flows from the South Atlantic to the North Atlantic, partly through the Caribbean Sea and Gulf of Mexico. Much of this poleward flow is powered by the Gulf Stream, which skirts and meanders northward off the US East Coast. Once back in the cold high northern latitudes, these surface waters lose heat to the atmosphere, increase in salinity, and the cycle of flow begins again.

The oceans' great conveyor belt brings heat from the tropics to the poles, helps to reoxygenate the oceans, distributes carbon and nutrients, and affects regional climates, the ocean food web, fisheries, storms, sea level rise, and more. A complete cycle is thought to take about a thousand years. In the context of today's warming, the big question is: Is the input of increased quantities of meltwater in the Arctic causing the oceans' density-driven circulation to slow, and could it eventually power down all together? This is a topic of intense interest. Geologic evidence suggests that in the past, slowing of the Great Ocean Conveyor Belt may have triggered abrupt and dramatic climate change.

Paleoclimate data from proxies such as ice cores, tree rings, and marine sediments indicate that at the end of Earth's last ice age a series of rapid climate shifts occurred. Ice cores from Greenland in particular show that regional temperatures swung dramatically within mere decades. For instance, around 12,800 years ago, when the Earth was gradually warming out of a big freeze, a rapid shift occurred, and the planet plummeted back into cold—at least in the Northern Hemisphere. Data suggest that the high northern latitudes became cold and dry with increased winds. A winter deep freeze hit the North Atlantic, Asian monsoons weakened, and tropical rainfall moved south. But the changes were not evenly distributed across the globe. When the quick chill hit the northern climes, Antarctica continued to warm, at least for a while. Then, a little over a thousand years later, the Northern Hemisphere swung back to warmth, with temperatures rising some 8 to 16°C in just a decade.

It appears that such abrupt changes were not unusual during the Earth's last ice age—the planet was hit with a series of large, rapid, and widespread shifts in climate. In fact, as meteorologist Jeffrey Masters explains it, during the past 110,000 years, there have been at least twenty abrupt climate changes. The only period in which there has been a relatively stable climate during the last 110,000 years, he points out, is the 11,000 years of modern climate.

What causes the Earth's dramatic and abrupt swings in climate? In several instances, the rapid cooling has coincided with enormous inputs of freshwater from glacial lake discharges or giant armadas of icebergs released into the North Atlantic. The predominant theory is that the addition of freshwater to the North Atlantic or at higher latitudes causes the oceans' big flow to slow, leading to radical cooling in the Northern Hemisphere. But the processes involved, the timing, and connections to asynchronous variations in the Antarctic are not well understood.

Hypothetically, with the addition of enough freshwater at high northern latitudes, where increasingly dense seawater normally sinks, deep-water formation could be reduced or stopped altogether. If this were to happen, it would back up and slow the rest of the ocean conveyor belt. Warm, northward-flowing surface water in the North Atlantic Ocean (the Gulf Stream) would then have nowhere to go, so less heat would be transported to higher latitudes. Northern Europe would cool dramatically, while the southeast coast of the US would warm and experience higher sea level (water expands when it is warmer). It is such an intriguing and dramatic cascade of events that even moviemakers like the concept, as dramatized in *The Day after Tomorrow*. It is important to note that a slowdown or halt of the ocean's overturning circulation isn't expected to result in catastrophic instant-freeze hurricanes or tornadoes as depicted in the movie. And thankfully, it also appears there is a restart mechanism.

So are the large amounts of meltwater and ice being released into the Arctic today, including from Greenland, slowing the oceans' overturning flow? Could the high-latitude driving force behind density-driven flow in the sea shut down completely in the not-so-distant future?

With time and new oceanographic tools at their disposal, scientists are beginning to identify variations in the oceans' great flow. Beginning around 1975, data indicate that the density-driven circulation in the North Atlantic weakened. For about two decades, transport was sluggish. It may have been the first such ocean slowdown in the last thousand years. After that, the system seems to have rebounded until about 2009/2010, when another slowdown occurred. The following winter in the United Kingdom was especially snowy and the chilliest since 1890. At the same time, along the northeastern US coast, sea level rose some 12 centimeters. Were these the result of the ocean's lagging flow? Flow in the North Atlantic then appears to have again strengthened. Some reports now suggest that overall, over the last decade, the ocean's great flow is weakening.

The global circulation of the ocean is in fact much more complex than is depicted in the conveyor belt model. Variations on a daily, weekly, annual, and decadal scale are influenced by complex interactions involving gravity and wind-driven flows, heat flux, precipitation, shifts in large-scale wind patterns, and processes of mixing. Changes to highly reflective ice, which can insulate the underlying ocean from heat loss, factor in as well. The truth is there's still *a lot* we don't know. Scientists are not even sure where meltwater from Greenland actually goes and how it gets there.

Because our data is limited in time and space, it is too soon to tell if the oceans' great and complicated pattern of flow is really slowing down, if at some point it could halt altogether, and if it did, what that would mean for the ocean, land, and people. A fuller and lengthier data set is needed. Again, stand by.

Other theories to explain past abrupt changes in climate focus on atmospheric dynamics, including variations in the El Niño–La Niña oscillation, the shifting position of tropical moisture and winds (the Inter Tropical Convergence Zone), and the influence of sea ice stability as well as high-elevation ice sheets. But it's a bit like the old chicken-and-egg conundrum. Which comes first, an atmospheric alteration or changes in ocean flow?

Because the ocean has a great capacity to absorb heat and ocean circulation takes time, there may be a long lag time even after a trigger point or threshold has been passed. Scientists estimate that since 1995, the ocean has absorbed 93 percent of the excess heat due to global warming. Are the impacts of global warming being delayed or dampened by the ocean's ability to absorb heat? When that capacity for buffering is exhausted, will the ocean's flow slow and sea level be more like it was 125,000 years ago— 5 meters higher than today? It's another very big wish-we-knew.

Scientists, young and old, are focusing their efforts on better understanding how the Earth responds to warming. They are also working diligently to identify and predict the impacts.

Extreme Weather Events

We can't necessarily say that extreme events are due to global warming, but it is highly likely that their extreme impacts would not have occurred without global warming. –**Kevin Trenberth**, National Center for Atmospheric Research

Extreme weather events are in the news nearly every day now: unprecedented flooding, intense and destructive wildfires, record-setting temperatures, severe drought, powerful storms. These events are not only dangerous and at times immensely destructive; they are also extremely costly. In 2017 there were sixteen weather and climate disasters in the United States that cost at least a billion dollars each. It was by far the costliest year on record, with cumulative costs topping $300 billion. Is climate change causing more extreme weather, and if so, will it get worse in the future?

For many years, scientists were apt to say it was too soon to tell or that the influence of climate change could not be attributed to individual events. Things have changed, however, particularly over the last decade or so. Scientists are now actively trying to determine the probability of extreme weather events occurring with or without the influence of climate change. Some researchers consider this a whole new field of science. Questions about extreme events focus mainly on their frequency, intensity, spatial extent, duration, and timing.

One perspective is that high-impact weather events are not being caused by climate change, but their impacts are becoming more extreme because of it. Several examples illustrate how climate change can exacerbate the impacts of a weather event. In a warming climate, more moisture is available in the atmosphere; this can fuel more intense and longer-lasting rain or snow events. The 2017 flooding caused by Hurricane Harvey, and the infamous Snowmaggedon that struck Washington, DC, in February 2010, illustrate how increased atmospheric moisture can cause impacts to go from bad to extreme.

In Hurricane Harvey, available moisture in the atmosphere from an unusually warm Gulf of Mexico combined with a lingering storm to produce catastrophic flooding (more in chapter 4). In Snowmaggedon, back-to-back storms dumped over a meter of snow on the US capital, causing hundreds of accidents and flight cancellations, and leaving hundreds of thousands of people without power. A February snowstorm is not out of the norm for the Washington, DC, area, but in this case above-normal sea surface temperatures in the nearby tropical Atlantic supplied an abundance of moisture to the atmosphere. With an extraordinary amount of moisture flowing into the storm, snowfall went from picturesque to disastrous.

Superstorm Sandy in 2012 also illustrates how climate change leads to more extreme impacts. On its track northward, Sandy passed over unusually warm ocean waters, fueling its strength and providing moisture for its intense precipitation (again, more in chapter 4). Sea level rise boosted Sandy's storm surge some 19 centimeters, also exacerbating the impacts. And Superstorm

Sandy's unusual westward motion just before landfall may have been a result of or influenced by climate change. Research now shows that changes in the Arctic are modifying the behavior of the jet stream, causing large meanders and slowing. Such behavior is thought to create instability and longer-lasting or more extreme weather events as in heat waves, flooding, or drought, and may have helped steer Sandy onto the coast. Data, as explained by Penn State climatologist Michael Mann, now detectably links the slowing and meandering of the jet stream to human activities, and is a reminder that the impacts of human-induced climate change are no longer subtle.

In late 2018, another extreme weather-related event occurred—one of the most costly and destructive wildfires in history. Within hours of flaring up, the Camp wildfire in California scorched thousands of acres. It moved faster than people could flee. Thousands of homes, cars, and buildings were destroyed. At least eighty-five people were killed and the town of Paradise was left with little but charred remains. Insured losses are estimated at more than $12 billion, with over $16 billion in total costs. Two other 2018 wildfires in California, the Carr and Woolsey burns, added to the heartbreaking and destructive season. Climate change isn't causing California's fire woes, but it is making them more extreme. Data indicate that due to climate change, California wildfires are more frequent, larger, and more destructive. Increased Santa Ana winds, higher temperatures, and lower humidity create conditions that lead to wildfires that can spread terrifyingly fast.

We can now say with confidence that due to climate change, the impacts of weather and weather-related events are becoming more extreme. As the climate continues to warm, will these impacts get even worse, and if so, how much worse? In droughts, will there be increased drying, heating, heat waves, and wildfire risks? In rain or snow events, will there be more moisture available in the atmosphere, making storms even more intense or longer lasting? Will wildfires become more widespread, intense, and destructive? Will floods occur in more or different places and be more frequent, more severe, and more persistent? A recent study indicates that the average number of tornados in large events or outbreaks has grown by 40 percent since 1954. Is this also a consequence of climate change? Whether we like it or not, the answers to these questions are headed our way.

What about secondary impacts? For instance, if more rain causes higher runoff or river flow, will this wash an increasing nutrient load into our lakes, rivers, and coasts fueling larger and more frequent algal blooms and an increasing number of dead zones? Some of the impacts of extreme weather due to climate change may also fall within the category of unknown unknowns—consequences we have yet to recognize.

Ocean Life

Some 71 percent—nearly three-quarters—of the planet is blue, blanketed by the sea. In the coming years, much of the Earth's response to climate change will happen in the ocean. As mentioned previously, the ocean has absorbed 93 percent of the excess heat produced by climate change. This finding is based in part on data coming from the long-running Argo float program. Thousands of automated Argo buoys now travel the world's oceans at depth (around 1,000 meters). They are programmed to drift with the water's flow, recording temperature and salinity, and to return periodically to the surface to transmit their data via a satellite link. Data from the Argo buoys has been and will be critical to solving some big science wish-we-knews in the sea.

But it doesn't take a scientist or thousands of high-tech buoys to recognize some of the impacts of climate change in the ocean. Species are expanding or altering their range, showing up in places where they've never before been seen. For example, American lobsters have marched north, king crabs have invaded Antarctica, and bottlenose dolphins have been sighted off British Columbia. Large shifts are also occurring in fish and plankton populations. And while the number of jellyfish and cephalopods in the sea appear to be on the rise, other organisms, such as Adélie penguins, polar bears, and beluga whales, are on the decline. Anomalously high temperatures in the North Pacific may also have triggered an infectious disease causing the worst mass wasting of sea stars ever recorded.

While scientists try to assess all of the changes taking place, they are also trying to predict what will happen next. It's important not just for the ocean and marine food web but for humans as well. To say we rely heavily on the sea seems an understatement. In the ocean, coral reefs may sadly be the bellwether of things to come for life in the sea and on the planet.

Coral Reefs

Coral reefs are among the most biologically diverse habitats on the planet. Covering less than 1 percent of the seafloor, they harbor thousands of marine species. They also support important fisheries, create enormous and valuable tourism and recreation industries, are a source for new pharmaceuticals, and protect coastal communities from storm waves, flooding, and tsunamis. On an annual basis, the worth of coral reefs is estimated at a trillion dollars.

Climate change poses no less than a triple whammy of threats to coral reefs: rising seawater temperatures, increasing ocean acidification, and accelerated sea level rise. This is on top of existing pressures such as overfishing, pollution, invasive species, and development. Events beginning in 2014 and continuing through 2016 are illustrative of the alarming coral reef crisis now occurring in the sea.

In 2015 and 2016 a strong El Niño combined with a warming ocean to cause massive bleaching of corals worldwide. According to the National Oceanic and Atmospheric Administration (NOAA) it was the longest, most widespread, and possibly most damaging episode of coral bleaching ever recorded, affecting reefs across the world, from the Caribbean and Florida to Japan and Hawaii, from Kiribati, American Samoa, Fiji, and Papua New Guinea in the Pacific Ocean, to the Maldives and Réunion in the Indian Ocean, to the southern Red Sea. In Kiribati, surveys found 80 percent of the corals dead and 15 percent bleached. Australia's Great Barrier Reef was hit particularly hard by warmer than average seawater temperatures. In its northern section the event was catastrophic, killing 50 percent of the reefs affected. Three such bleaching events, in 1998, 2002, and 2016, have now impacted some 85 percent of the Great Barrier Reef. While El Niños have been occurring for hundreds of years, a background of increased ocean temperature due to climate change has swung the hammer on the world's coral reefs.

What is coral bleaching? Within their tissues, reef-building corals host an algae partner. The algae use the coral's waste products as nutrients and fuel for photosynthesis, while the coral gets an efficient waste removal process and oxygen production system. These processes facilitate the coral's ability to produce a calcium carbonate (limestone) skeleton and to build reefs. Prolonged seawater temperatures, high or low, outside of a coral's optimal range can cause stress and the expulsion of its symbiotic algae. With the loss of its algae, a coral's tissues become transparent and its white rocky skeleton visible—it appears white or bleached (plate 5). Some corals can survive bleaching if the stress is short-lived and the algae partners return. But if the stress is acute enough or prolonged, coral bleaching is often fatal.

The big unknowns here are worrying. Is the pace of warming too fast for corals and their algae to recover and adapt to higher seawater temperatures? Is this the start of the demise of corals globally? Corals grow slowly, and reefs need time to recover. High seawater temperatures can also make corals more susceptible to disease, like the one that has recently spread throughout the Florida Keys and into the Caribbean, further decimating the region's reefs. If the conditions that cause bleaching occur again sooner rather than later, it may mean a death knell for reefs as we know them and create an economic, food, and job loss disaster for human society.

Unfortunately, it's not just rising seawater temperatures that are a problem for coral reefs. If sea level rises too quickly, it can outpace a reef's ability to keep up or grow upward. The water may become too deep for algae within the coral to obtain adequate sunlight for photosynthesis. And with increasing carbon dioxide in the atmosphere, more has been absorbed into the ocean, lowering its pH (in other words, increasing its acidity). For organisms that create skeletons or shells of calcium carbonate, pH matters: increased acidity may impede their ability to create their rocky homes. Chris Langdon at the University of Miami has been exposing corals to varying temperatures and acidity levels. To his surprise, he's found great variability within and between corals in how they react to changes in temperature and pH. There's some hope that a few supercorals exist, able to withstand and adapt more quickly than expected. While large-scale reefs may disappear or change dramatically (becoming dominated by algae or sponge), some corals will hopefully survive.

But skeleton building by coral is not the only thing affected by increasingly acidity in the ocean. The biological functioning of marine creatures—their respiration, breeding, and growth—is also threatened. The shellfish industry is already on notice. In Oregon, oyster production at the Whiskey Creek Shellfish Hatchery nearly crashed because the ocean water pumped into the facility became too acidic for its oyster larvae. The company must now spend a million dollars a year to add more alkaline water to the incoming ocean water. The overall consequences of an increasingly acidic ocean remain unknown, but it doesn't look good for many of the species that live in the sea.

Dead Zones

OFF THE COAST IN ONE OF THE MORE THAN FOUR HUNDRED DEAD ZONES. For the fish and other organisms in the area, it didn't matter if the decline in oxygen was gradual or abrupt. It didn't matter the cause. For these creatures, the only thing that mattered was breathing, and that was getting hard to do. The fish began to gulp, trying to pass as much water as possible over their gills. The oxygen levels in the water then dropped even lower, dangerously lower, and instinct kicked in—at least for the fish and shrimp that were still strong enough. They swam away. Other creatures lingered, getting weaker and weaker. On the seafloor, crabs, worms, and clams couldn't escape. Once the oxygen became too low they were doomed. Soon the ocean floor was littered with the dead and dying.

A dead or hypoxic zone occurs when oxygen concentrations in the ocean are too low to support life; more than four hundred have been documented across the world's oceans. The poster child for dead zones forms each summer in the Gulf of Mexico off the outlet of the Mississippi River. In the summer of 2017, the Gulf of Mexico dead zone was the size of New Jersey—over 20,000 square kilometers. In 2018 it was smaller, about average—about the size of Connecticut.

The cause of dead zones is not an unknown—they are the result of excess nutrients, in the form of nitrogen or phosphorous, typically from fertilizer and other wastes washing off the land into rivers, lakes, and the sea. The Mississippi River has one of the largest drainage basins in the world. As the river flows into the Gulf of Mexico, it brings with it a whole lot of nitrogen. When conditions are right, that nitrogen spurs the rapid growth of algae, and thus a bloom ensues.

How does an algae bloom create a dead zone? With an abundance of algae at the surface, sunlight is blocked from the photosynthesizing organisms below and the production of oxygen shuts down. As the algae dies, it sinks to the bottom. What oxygen remains is further used up in decomposition and concentrations go from low to deadly. Dead zones are more prevalent in summer months because warmer surface water can create a low-density cap and prevent vertical mixing (which brings in oxygen).

As global temperatures rise, so too may the number of dead zones. Warmer water holds less gas (including oxygen) in solution, increased warmth at the surface inhibits mixing, and increased rainfall means more runoff, which in turn means more nutrients entering the sea.

Oxygen levels also appear to be changing on a larger scale. Research indicates that in the global ocean, between 1960 and 2010, a small but widespread decline in oxygen content occurred. Unfortunately, there is a lack of long-term data on the ocean's overall oxygen content, so it is difficult to delineate the impacts of climate change. It is another challenging and important wish-we-knew that researchers are working on.

Consequences of Change

There is a shocking, unreported, fundamental change coming to the habitability of many parts of the planet. –**Peter Gleick**, cofounder of the Pacific Institute

Across the world, every day, the impacts of climate change are being played out, at the planet's northern and southern extremes as well as in temperate and tropical forests, in wetlands and marshes, in rivers and lakes, across deserts and plains, on mountain slopes and along our coasts. Rapidly changing conditions are affecting wildlife species from the large and charismatic to the smallest and unseen. Entire ecosystems and food webs are changing, and dramatic population declines or mass extinctions are reportedly occurring across the globe.

The impacts and toll of climate change on human society are also increasingly being felt. Extreme weather events made worse by climate change are costing billions of dollars annually—a cost that is expected to rise in the years to come. The US Department of Defense has officially stated that climate change poses a serious threat to national security. Changes in precipitation, heat, and drought are significantly impacting farming, food supplies, human health, safety, and disaster management, recovery, and response. Access to fresh water is already a critical issue in many parts of the world; in a warming world the problem is expected to worsen and become more widespread. The spread of mosquito-borne diseases is also projected to increase. And with significant sea level rise the amount of livable space on the planet is going to shrink.

Industry is and will feel the brunt of our changing climate: from transportation and construction to insurance, health care, fisheries, tourism, and real estate. And climate change has the potential to cause a refugee crisis unprecedented in modern times. One study suggests that by 2050 nearly a third of the world's land surfaces will turn arid and become desertlike. Large swaths of Asia, Europe, Africa, and southern Australia will be impacted, along with the more than 1.5 billion people in those regions. Some posit that mass migrations caused in part by climate change are already occurring: out of low-lying, land-impoverished Bangledesh, from Saharan and Sub-Saharan Africa, and in Syria where, before the war, three years of drought resulted in over eight hundred thousand people, mostly farmers, losing their livelihoods.

The data are clear and present. There is no longer any credible scientific debate. Some 97 percent of climate scientists agree that climate change is happening and the principal cause is the increase of carbon dioxide in the atmosphere due to human activities. Yes, we still need to determine how much and how fast the ice sheets are melting and how much that will contribute to rising sea level. Yes, we need an improved estimate of how much of the world's permafrost will thaw and how soon, and to determine if the ocean's great global flow is in fact slowing down. And all across the planet, we need a better understanding of the complex and closely linked interactions between the ocean, atmosphere, ice, and land. But the world is already changing, and faster than previously predicted. Extreme weather events are causing costly and tragic losses across the world. So while it is imperative that we continue to invest in science to better our understanding of climate change and more accurately predict its impacts, we absolutely know enough to warrant action now.

Maybe the biggest unknown is this: what will we do? Will emissions of carbon dioxide and other greenhouse gases be reduced, hold steady, or escalate? Will policies be put in place to promote the development of carbon-capturing technologies and the use of alternative energies? People were ingenious enough to discover and take advantage of fossil fuels; we are equally capable now of fixing the problems burning carbon has led to. But it is going to take political will and fundamental changes to society and our way of life.

Any attempt to address and mitigate climate change will have to be multifaceted. We must reduce emissions. We need to conserve. And we need policies that do away with subsidies making fossil fuels artificially cheap, create incentives to assist in transitioning from fossil fuels to alternative energies, and invest in innovation. But reducing carbon emissions will not be enough. We must also find ways to remove carbon dioxide from the atmosphere. Project Drawdown offers a hundred effective ways to reduce emissions and sequester carbon from the atmosphere (see the book edited by Paul Hawken in the list of readings and references at the back of this book). From wind, solar, and geothermal power to a plant-based diet, food waste reduction, improved agricultural practices, education, family planning, and infrastructure changes, it is a hopeful look at what can be done across the globe. Many of these practices can be pursued independent of the skeptics and those whose agendas favor short-term profit over the long-term welfare of the world's citizens. Individuals can make a difference, and are doing so, by changing what is consumed, bought, built, grown, and invested in. The time to act is now, not later.

Based on the scientific data, the Earth is warming unnaturally fast, and the consequences for the near future are nothing short of alarming. Questions and unknowns remain, but as the current residents and caretakers of this planet we need to do as much as possible so that future generations don't look back and wonder why we didn't act sooner.

2

Volcanoes

For years, a strange landscape found near volcanoes mystified scientists. Neither historical records nor geologic deposits from previous volcanic eruptions explained the scatterings of variously sized, slightly peaked or rounded hills referred to as hummocky terrain. Without knowing how this landscape formed, there was no way to predict the extreme danger it forewarns.

MOUNT ST. HELENS, WASHINGTON, 1980. For more than 120 years the mountain lay quiet, a snow-covered cone-shaped peak surrounded by huge swaths of old-growth forest. At the base of its northern slopes lay the cool, clear waters of Spirit Lake. Abundant elk and deer grazed among the trees while in the lake swam steelhead and salmon. The setting was idyllic; a wonderful place to live and visit, offering serene views and ample opportunities to fish, camp, hike, and swim. Unbeknownst to many, Indian legend and buried ancient rocks spoke of the mountain's violent temper—the danger next door.

In March 1980, a series of relatively small earthquakes rattle Mount St. Helens. Small avalanches of snow and ice cascade off the mountain's peak. The tremors increase in size and frequency, and soon a steam-powered blast rocks the summit. Steam and ash shoot 2,000 meters skyward, and a fractured 76-meter-wide crater opens at the volcano's peak. Over the next several days more steam blasts occur, and the crater gets wider and deeper. A rhythmic ground motion known as volcanic or harmonic tremor is detected, indicating that underground magma or molten rock is on the move. Only one seismometer sits directly atop Mount St. Helens. At the US Geological Survey, Robert Tilling is in charge. He doesn't have the funding to send scientists and equipment to Mount St. Helens. But he knows what's at stake, so he pulls out a credit card and sends them anyway. Scientists are rushed in to assess and monitor the volcano. Is it the beginning of a major eruption or just a period of unrest? It is a big unknown—and a question no one could answer.

For the next month and a half, Mount St. Helens is only slightly and intermittently restless. Ash and steam emissions decline. Fewer earthquakes occur each day, but the average size or magnitude increases. Scientists monitor the sulfur dioxide emissions—a sign of magma at shallow depths releasing gas. The readings are negative.

On May 7 a small steam explosion occurs, which is followed by a series of intermittent blasts. Then, on the mountain's northern flank a bulge emerges. Beneath it, earthquakes shake the mountain. The rocky bulge soon swells, growing about 1.5 meters a day. Meanwhile, the ground behind the bulge appears to deflate. The experts on-site are concerned. With the equipment available, they anxiously keep watch on conditions, recording the rate of the bulge's growth, gas emissions, and earthquakes. If an eruption is going to happen, they expect ground deformation to accelerate and earthquake activity to intensify. But the rate of bulge growth remains steady, and the earthquakes show no signs of change.

On the morning of Sunday, May 18, 1980, the sky over Mount St. Helens is clear and visibility unusually good. It is a postcard-perfect day. From his observation post some 10 kilometers north of the mountain, USGS volcanologist David Johnston reports that all is normal. Measurements of seismic activity, gas emissions, ground temperature, and deformation show nothing un-usual. Then at 8:32 a.m., a 5.1 magnitude earthquake rocks the volcano. Seconds later, the unthinkable happens. The bulge on Mount St. Helen's northern flank collapses and releases the vol-cano's fury.

Two gigantic landslides accelerate down the mountain's north-ern slopes. Enormous boulders, sand, and blocks of rock literally slide off Mount St. Helens and race into Spirit Lake, going up and over 300-meter-high ridges. At the summit, the removal of rock has uncorked what lay below, and a mixture of molten rock, gas, and water explodes up and laterally to the north. A searing flow of ash, gas, and rock races off the mountain at a speed of at least 480 kilometers per hour—it is a deadly volcanic phenomenon known as a pyroclastic flow. Ash and gas rise thousands of meters before collapsing and adding to the debris plummeting off Mount St. Helens. Trees are laid down like toothpicks (figure 2.1a). Within minutes, 600 square kilometers become an ash-covered waste-land. The swelling ash cloud and surge soon enshroud most of the mountain. Explosions are heard hundreds of kilometers away, though up close the cataclysm is silent.

Another earthquake rocks Mount St. Helens—it's a 5.0 mag-nitude. Explosions burst to the east and west. A third huge av-alanche takes what remains of the mountain's summit with it. Another, larger pyroclastic flow then pours off the volcano and speeds across ridges and hills and into valleys. A deadly cloud hun-dreds of meters thick now hugs the land.

But the eruption isn't over. From the blown-off top of Mount St. Helens, explosions launch huge amounts of ash and gas sky-ward (figure 2.1b). Within ten minutes, a roiling mushroom cloud reaches some 19 kilometers high. Lightning bolts flash within the cloud and spark forest fires. Winds spread sand, ash, and tree lit-ter east. Day turns to night for hundreds of kilometers. Ash even-tually enters the stratosphere and encircles the globe.

And still another deadly volcanic hazard looms. The eruption has melted the glacial ice and snow blanketing Mount St. Hel-ens. Soon volcanic mudflows, called lahars, pour down its slopes.

Figure 2.1. (a) Tree blowdown and Fawn Lake after 1980 eruption; note Mt. St. Helens in the background. Courtesy Lyn Topinka/UGSG. (b) Mount St. Helens, 1980 cataclysmic eruption. Courtesy USGS.

They mix with heated groundwater gushing from avalanche debris and race through valleys and over 70-meter-high hills. The North Fork Toutle River turns chocolate brown, becomes laden with huge logs, mud, and debris, and rises more than 6 meters above normal. Everything in its path, including camps, homes, and bridges, is destroyed. The largest mudflow travels some 120 kilometers, leaving utter devastation in its wake.

The 1980 eruption of Mount St. Helens was the worst volcanic disaster recorded in US history, killing fifty-seven people, injuring more, destroying some 250 homes, and causing losses estimated at more than a billion dollars. The event highlighted much of what was known about volcanoes at the time, but also what was unknown or unexpected. Several aspects of the eruption surprised even the experts. And even today, knowing exactly what a volcano is going to do, and when, remains a wish-we-knew.

A Previous Warning

In 1975 scientists Dwight Crandell, Donal Mullineaux, and Meyer Rubin published a paper suggesting that, within the Cascade Range, Mount St. Helens was the mostly likely volcano to erupt and that an eruption could occur "before the end of the 20th century." It was a bold statement. Rarely do scientists make such forecasts. A few years later, they expanded on the hazards posed by an eruption of Mount St. Helens. Their report and hazards map were limited, based only on what they could glean from relatively recent geologic deposits (within the past four thousand years). But the tale those rocks told was ominous.

The evidence suggested that Mount St. Helens had a history of intermittent dormancy followed by eruptions that varied in intensity, style, and composition—from non-explosive flows of lava to violent ash ejections and dangerous pyroclastic flows and mudflows. Based on the data, there was no way to know what type of eruption would occur next, when it would happen, the sequence and speed of events, or the scope and direction of the impacts. At the time, most people also equated volcanic eruptions with picturesque, slow-moving lava flows.

During the 1980 eruption, scientists were especially surprised by the massive debris avalanche and accompanying lateral surge or blast. At the time, only a few people considered it a possibility. A Russian scientist had previously suggested that a lateral blast occurred during the 1956 eruption of the Bezymianny volcano on Kamchatka Peninsula, north of Japan, but he had not directly witnessed the eruption. In a technical report, scientist Barry Voight suggested the possibility of a gigantic landslide at Mount St. Helens and of its triggering explosions—but no mention of this was made in discussions before the eruption, and his report was received the day after.

Tragically, it was the debris avalanche and directed surge that caused most of the fatalities, including the death of volcanologist David Johnston. Vast swaths of forest were laid flat and Spirit Lake was filled with debris, raising its bottom some 90 meters. The debris avalanche also solved a geologic mystery—that of hummocky terrain. The collapse of the bulge, massive debris avalanches, and lateral blast during the eruption left behind a familiar looking landscape—scattered, slightly peaked hills and mounds, some nearly 50 meters high.

Hummocky terrain, like that produced in the 1980 Mount St. Helens eruption, has now been recognized near more than two hundred volcanoes worldwide, including California's Mount Shasta. This means each of these volcanoes has and may again produce powerful explosive eruptions, with huge debris avalanches and lateral surges. A distinctive U-shaped crater, as left behind at Mount St. Helens, is another clue to a volcano's potential for similarly violent events.

The disastrous 1980 eruption was a watershed event in the science and monitoring of volcanoes and in the response to volcanic activity. Across the globe, people became more aware of the very real hazards posed by volcanoes. The Mount St. Helens event exposed gaps in our understanding of volcanoes, a lack of sufficient monitoring, and the need for improved coordination during eruptions between local authorities, emergency response officials, and the scientific community. Firsthand observations at Mount St. Helens also revealed how rapidly a multiphase eruption can progress and that even distant structures and communities can be at risk.

In hindsight it is easy to question why more wasn't done to prepare when, five years earlier, scientists had warned of the likelihood and potential danger posed by an eruption at Mount St. Helens. Looking back on the eruption some twenty years later, geologist Robert Tilling noted that requests by the US Geological Survey for increased appropriations for research on volcanic hazards, particularly within the Cascades, had been repeatedly rejected. Predictably, after the Mount St. Helens disaster, funding to the USGS was increased and the Cascades Volcano Observatory was established. Investments in science and monitoring related to volcanoes and other hazards are all too often unavailable until a crisis occurs. This is a story that tragically repeats itself over and over.

Volcanoes such as Mount St. Helens exemplify our love-hate relationship with Mother Earth. On the plus side, volcanic eruptions have created most of the land we live on (some 80 percent of Earth's crust), including dramatic landscapes that support tourism and recreation. Ash provides fertile soil for agriculture, and in some areas, volcanic heat is a source of clean, natural energy. On the minus side, volcanic eruptions can cause catastrophic loss of life, lay waste to vast tracts of land, and destroy infrastructure and other property. So, while we enjoy the benefits of our beloved and majestic mountains, we must also understand and prepare for the dangers they harbor.

Today, more than half a billion people live dangerously close to the world's fifteen hundred active volcanoes. Each year there are a hundred or so volcanic eruptions. As human populations grow, so do the risks posed by volcanic events. Fortunately, with research and advances in technology, our understanding of volcanoes has progressed significantly. Much of what we have learned, however, is how much we still don't know; previous generalizations may need to be discarded and textbooks rewritten. Across the globe, scientists are working hard to fill in the volcano wish-we-knews, especially as they pertain to our ability to forecast eruptions and save lives.

Volcanoes: The Known

Where and why volcanoes occur is better understood today than ever before. In the past, information came from historical records, geologic deposits, and observations. Today, we have a growing array of additional scientific tools to study and monitor volcanoes. But the greatest leap in our understanding of where volcanoes are located and why, came with the scientific revolution known as plate tectonics.

Early on, many scientists thought the idea of rigid plates moving across the Earth's surface was simply bonkers. In the 1960s, when geologist Roberts Yeats was a graduate student, he learned of an idea called "new global tectonics." Yeats and his friends came to realize that much of the stuff they'd been taught in school was wrong and that their professor Harry Wheeler, who they'd thought was a nutcase, had been right all along. Plate tectonics is now accepted as fact and helps to explain much about the Earth that previously fell into the category of unknown. Following are a few plate tectonic basics as they pertain to volcanoes. For more on how the concept developed and was confirmed, see the list of readings and references at the back of this book.

Much like a giant jigsaw puzzle, the Earth's surface is divided into about fourteen relatively thin, rigid, and irregularly shaped pieces or tectonic plates. Relative to the rest of the Earth, these plates are razor-thin, like the skin of an apple or the bumpy surface of a basketball. Each moves independently at about the speed fingernails grow, averaging around 2 centimeters per year. The plates are topped by relatively thin oceanic crust (on average about 7 kilometers thick) or thicker continental crust (about 35 kilometers thick). Underlying the crust is a layer of the Earth's interior, or mantle. The tectonic plates sit on and move atop the asthenosphere—a partially molten layer of the mantle. Imagine the layer of sticky, somewhat fluid caramel that lies below the hard chocolate shell of a Milky Way bar.

Active volcanoes occur principally in three plate tectonic settings: where two plates diverge (move away from one another), where two plates converge (come together), and at hotspots. Most of Earth's volcanoes are actually underwater, associated with the 65,000-kilometer-long mid-ocean ridge system. Here, tectonic plates are diverging or spreading apart and magma rises to the surface to form an undersea chain of mountains.

Magma is essentially a mix of hot fluid or melt, crystals, and bubbles. The chemistry and proportion of each component varies. At mid-ocean ridges and hotspots, magma tends to be silica-poor and rich in iron and magnesium (basaltic). Because of its low silica content, this type of magma (or lava) tends to be less viscous (more fluid) than silica-rich magmas (rhyolitic). Gas escapes more readily from silica-poor magma, so eruptions are less explosive and more fluid (think Hawaii and broad, shield-shaped volcanoes). This type of magma or lava is the most common on Earth. Silica-rich magmas contain more potassium and sodium, tend to form thick blocky lava flows or domes, and erupt more explosively (think Mount St. Helens and cone-shaped strato-volcanoes). For the purposes of this book, "magma" and "lava" will be used as general terms to describe both the silica-rich and silica-poor varieties.

Sometimes, when two tectonic plates come together, or converge, one plate sinks or is dragged down beneath the other. This is called subduction, and where it happens is known as a subduction zone (figure 2.2). A deep-sea trench is the surface expression of an underlying subduction zone. Subduction zones create and fuel some of the most explosive volcanoes on the planet, such as in the infamous Pacific Ring of Fire, which includes volcanoes in the US Pacific Northwest, Japan, Indonesia, Mexico, and South America.

Here's how we think subduction zones create volcanoes: As one tectonic plate sinks beneath the overriding plate (due to gravity dragging it down and convection in the underlying, somewhat fluid asthenosphere), it is subjected to increasing temperature and pressure. At a certain depth, high pressure causes water in the descending slab to be released. The water then gradually seeps into the overlying hot rock or mantle and lowers its melting temperature. Magma forms that is less dense than the surrounding rock. It then rises buoyantly toward the surface, passing through

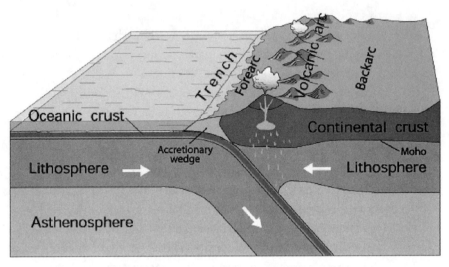

Figure 2.2. Subduction zone and the creation of volcanoes. Courtesy USGS.

existing fractures or breaking rock as it goes. On its way up, the magma may settle within cracks or larger spaces that act as underground reservoirs. An eruption occurs when enough magma is present and sufficient pressure is exerted from below or the overlying weight is removed from above. Over time and with multiple eruptions a volcano grows. A linear chain of volcanoes, such as the South American Andes, is often found parallel to and behind a subduction zone.

The third plate tectonic setting in which volcanoes occur is a hotspot. Hotspots are typically found in the middle of tectonic plates, with the exception of Iceland, which also sits on a mid-ocean ridge where the plates are diverging. Hotspots are locations where, for some unknown reason, deep in the Earth, possibly at the outer core-mantle boundary, temperatures are significantly higher and heating causes magma to rise buoyantly toward the surface. These locations appear to be stationary or in some cases may migrate slowly over time. Some twenty-five active and old hotspots have been identified, including Hawaii, the Galápagos, the Canary Islands, and Yellowstone. Much about hotspots falls into the unknown or wish-we-knew category—notably, what determines their location, and what turns them on and off? In fact, what turns the supply of magma on and off, and what controls the rate of supply, are big questions for all volcanoes.

With the understanding brought by plate tectonics, we can confidently say why most volcanoes are where they are. Recent eruptions provide another important means by which scientists learn about volcanoes. With each event comes information about what happened before, during, and after. Unfortunately, the knowledge sometimes comes with a tragically high price. Three eruptions in relatively recent history have been particularly important in advancing volcano science, teaching us hard lessons about the dangers posed and highlighting some critical wish-we-knews. The 1980 eruption of Mount St. Helens is the first of these influential events; the other two are Nevado del Ruiz in 1985 and Mount Pinatubo in 1991.

Nevado del Ruiz, Colombia

Located 120 kilometers west of Bogota, Nevado del Ruiz, at over 5,300 meters high, is one of Colombia's tallest peaks. *Nevado* means "snowy" in Spanish, and as its name suggests the volcano is capped by snow and ice year-round. Over the last ten thousand years, ten major eruptions have occurred at Nevado del Ruiz, but there have also been minor events. And here, even small eruptions can be extremely dangerous.

NOVEMBER 1984. High on the slopes of Nevado del Ruiz, mountain climbers report earthquakes and the presence of steam plumes. On the volcano's ice cap, ash and a veneer of sulfur are spotted. A landslide occurs.

Two months later, a geologist visits Nevado del Ruiz and finds a new small crater at the volcano's summit. But there's little expertise in the area or monitoring equipment to investigate further. Then, in February 1985, the only seismometer on the mountain fails. A call for help goes out. Volcanologists from Colombia and elsewhere respond and begin investigating. Experts confirm a high level of earthquake activity and suggest it could be a precursor to an eruption. But still there is no monitoring system in place on Nevado del Ruiz, nor are there people on-site with the needed expertise.

SEPTEMBER 1985. An explosion on Nevado del Ruiz blasts steam, ash, and blocks of rock into the air and onto the mountain's ice cap. Ash and rain fall as far away as 25 kilometers and an avalanche of ice, snow, and volcanic debris flows 27 kilometers down the mountain. The eruption lasts seven hours but has little impact on the surrounding communities.

For the remainder of the month, Nevado del Ruiz is sporadically restless. Then volcanic activity diminishes. Things seem to be quieting down. But volcano scientists, regional authorities, and emergency personnel are worried and suggest the public should be warned about the risk of avalanches should a new and larger eruption occur. Government officials and the media disagree, claiming that the volcano's activity is not dangerous. Their priority is to keep local communities calm. The local archbishop goes so far as to call the warnings "volcano terrorism," and a magazine warns of "real estate devaluation." Attempts to monitor the seismic activity on the volcano are hampered by a continued lack of equipment and expertise, and by failures of the few instruments available. Nonetheless, a preliminary hazards map is presented to authorities. Should Nevado del Ruiz erupt, evidence suggests a mudflow will threaten the residents of Armero, a village with some twenty-nine thousand residents 45 kilometers from the volcano's summit. However, if a timely warning is given, the people of Armero should have two hours to evacuate.

In mid-October, earthquake activity at the volcano increases and a small steam blast occurs. But still there is no evidence to suggest an eruption is imminent. Then, in early November, a swarm of high frequency earthquakes is detected, followed by continuous volcanic tremor (a type of earthquake that indicates magma is moving underground, like vibrations caused by water rushing through a pipe). On November 12, a geologist climbs to the volcano's summit to measure gas emissions. Again, he finds no indication of an impending eruption.

NOVEMBER 13, 1985. Just after 3:00 p.m., Nevado del Ruiz blasts ash and steam skyward. It is a relatively minor eruption, similar to the one the previous year. An hour later, ash begins falling on Armero. Rain turns the ash heavy as it blankets the town. A few more small explosions and tremors occur. By 7:30 p.m., Nevado del Ruiz has quieted down; authorities on the radio and a local priest reassure people that all is fine. Then, just after 9:00 p.m., two strong explosions rock the volcano and a series of searing pyroclastic flows surge across the ice cap and down the mountain's slopes. Another explosion blasts ash, gas, and glowing bombs high into the sky. Ash and debris from the eruption blow east, landing downwind as far as 100 kilometers away. Rain and ash again fall on Armero.

Meanwhile, atop Nevado del Ruiz, the mountain's namesake ice and snow are rapidly melting and earthquakes are causing unconsolidated water-laden sediments to liquefy. Mudflows race off the mountain and are channeled into the area's steep-sided river valleys. As the mudflows sweep downward, they coalesce and grow, entraining additional sediment, debris, and water. One mudflow pours into the town of Chinchina, destroying hundreds of homes and killing some eighteen hundred people. At about 11:00 p.m., another mudflow laden with boulders and debris races downstream—at the front is a wall of mud up to 40 meters tall. The flow roars into Armero. Over the next two hours successive surges plow over the town, burying it under 6 meters of mud and debris. Armero is all but obliterated. In a 1998 retrospective on the disaster, geologist Barry Voight described the scene as a "crypt sealed in lahar mud."

More than twenty-three thousand people perish and five thousand are injured in Armero. Those who survive are encased in a thick mixture of mud, trees, and debris, unable to move. Near the volcano, every road, bridge, communication line, power grid, and aqueduct is either damaged or destroyed. Seven thousand people are left homeless, livestock are killed, and crops destroyed. The cost of the eruption is estimated at more than a billion dollars.

The disaster wrought by the 1985 eruption of Nevado del Ruiz rocked Colombia and the science community. Days before the eruption, local authorities had been given a report on the hazards posed by the volcano. But the officials deemed the report "too alarming." They did not circulate it, nor, when ash began falling on their village, did they order an evacuation. Once the eruption began, some people tried to warn residents of the impending disaster, but communications were down and time was short. Just as important, the people of Armero did not believe that they were at risk and had no emergency plan in place should a warning be given. As predicted, there were two hours between the eruption and when the mudflow arrived in Armero—enough time for an evacuation. It was also discovered that Armero had been built unknowingly on the remains of a mudflow from a previous eruption of Nevado del Ruiz.

One of the hard lessons learned from the 1985 Nevado del Ruiz eruption was that with the right conditions, even relatively small volcanic events can cause disaster. And again, as at Mount St. Helens, precursor activity can be difficult to interpret—especially without a monitoring system in place early on to distinguish increased activity from typical background conditions.

The tragic consequences of the 1985 Nevado del Ruiz eruption also helped to change how Colombia and the world respond to volcanic unrest. After the event, the Volcano Disaster Assistance Program (VDAP) was established by the US Geological Survey and the US Agency for International Development's Office of Foreign Disaster Assistance. Now, when volcanic unrest occurs, governments can request the assistance of VDAP in monitoring, assessing hazards, generating eruption forecasts, and developing early-warning capabilities. The importance of the improved response by the international community and VDAP has been no better illustrated than when six years later one of the largest eruptions of the century occurred in the Philippines. The event was also another milestone opportunity for scientists to learn about volcanoes.

Mount Pinatubo, Philippines

On the small island of Luzon, for more than five hundred years, Mount Pinatubo slumbered quietly. In Filipino, *pinatubo* means "to grow." The land surrounding the mountain was densely forested and incredibly fertile (thanks probably to past eruptions), supporting widespread agriculture. In front of Mount Pinatubo lay a large expanse of flat ground, perfect for the giant runways of Clark Air Base. Not far to the south was Subic Bay Naval Station. By 1991 nearly a million people lived in the shadow of the mountain. Many didn't even realize it was a volcano.

JULY 1990. A major 7.8 magnitude earthquake rocks Luzon, killing sixteen hundred people. Two weeks later, a group of nuns visits the local office of the Philippine Institute of Volcanology and Seismology. They are worried about the shaking and opening of new steam vents on their mountain—Mount Pinatubo. A team sent to investigate finds no cause for concern. It is assumed that the recent earthquake shook up the volcano's plumbing and opened the new steam vents.

The following year the nuns return. They remain concerned about Mount Pinatubo's unrest. Another team is sent to the mountain, and this time they begin to detect small earthquakes originating northwest of the summit.

MID-APRIL 1991. Mount Pinatubo is now awake and restless. Each day, 80 to 130 earthquakes rattle the area around the volcano. Violent bursts of steam blast upward, and three large craters have opened on the mountain's northern flank. Experts in the region contact the US Geological Survey asking for help, and together they try to determine if the new activity is a sign of magma moving upward or simply the result of water-related (hydrothermal) processes. The region is well known for its geothermal activity.

With the help of the meteorologist and officers at Clark Air Base, and after wrestling with local bureaucracy, an official request for help is made to the Volcano Disaster Assistance Program. Funding for a response isn't available, but still a USGS team is sent to Luzon to work with experts at the Philippines Institute of Volcanology and Seismology. Together, they develop a strategy to address Mount Pinatubo's growing unrest with increased monitoring, education of the local community about potential impacts should an eruption occur, and an investigation into the volcano's eruptive history as recorded in the surrounding geologic deposits.

A startling discovery is soon made near the edge of Clark Air Base. In a 70-meter-high canyon wall, the scientists find layer upon layer of rock and sediment from past eruptions. Previous eruptions of Pinatubo weren't just explosive and big; they were enormous and pumped out pyroclastic flows and mudflows that extended even farther than those of the 1980 Mount St. Helens eruption. Further examination of the sediments, along with radiocarbon dating of charcoal bits, enables the scientists to estimate that Pinatubo has erupted about once every thousand years, give or take five hundred—with the last eruption being about six hundred years ago. Is the volcano primed and again ready to blow?

Over the next several months, teams battle the jungle vegetation, difficult terrain, and bureaucracy to set up additional seismic stations. They confirm that the earthquakes are centered some 5 kilometers to the northwest of Mount Pinatubo and about the same distance down. But they still can't determine if the tremors are related to the last big earthquake or to hydrothermal activity, or if they are a sign that magma is dangerously on the move.

In the meantime, geologist Chris Newhall and others begin work to educate people in the surrounding communities. One of their most effective communication tools is a video made by volcanologists and filmmakers Katia and Maurice Kraft. Seeing the destruction volcanoes have wreaked elsewhere, such as at Nevado del Ruiz, helps convince people of the very real dangers Pinatubo presents. The team also pushes local leaders, including those at the air base, to plan for evacuations if needed.

In May scientists rig a gas emissions detector to an air force helicopter. The instrument was originally designed to detect pollution in smokestack emissions. They fly in circles, low over Pinatubo. The survey reveals that the volcano is releasing 500 tons of sulfur dioxide a day. It isn't a huge amount to the scientists, but it does indicate that, below the volcano, magma is releasing gas fairly close to the surface; it isn't simply hot water.

The science team steps up its efforts to convince those living on and around Mount Pinatubo: the volcano is awake, stirring, and poses a very serious threat should it erupt. Still, some people see the scientists' research as esoteric or, worse, a cover story for financial scheming or a political agenda (sounds eerily similar to the arguments put forth by some climate change skeptics).

In late May 1991, seismometers pick up an unusual long-period earthquake. Scientists interpret it to mean that magma is moving underground. A tremor then occurs directly beneath some vents, and an explosive steam blast shakes the mountain. The experts remain unsure. Is this a change leading up to an eruption or just lingering unrest?

Early June brings new and changing data to the team monitoring Mount Pinatubo. Earthquakes continue to occur to the northwest, but now also cluster under the steam vents on the mountain. Grayer ash-laden steam shoots higher, up to 300 meters, above the summit. Analysis of the ash reveals that it comes from old rock, not new magma. Sulfur dioxide emissions then drop precipitously, from 5,000 to 260 tons per day. This worries the scientists. If gas is no longer freely escaping from the volcano, it could be building up below—pressurizing the mountain for a blast. Has rising magma formed a viscous plug or cork beneath the summit?

Soon a series of earthquakes shakes the area, including a long-period event again suggesting magma is on the move. People take more notice this time and become alert to the possibility of an impending eruption. The question of an evacuation hangs tensely in the air. Ash emissions grow, and then a lava dome is spotted on Mount Pinatubo's summit. Poking out of the east wall of the volcano is an ominous giant blob of gray rock.

Over the following days earthquake activity increases, an explosion occurs, and ash emissions continue. Then a defining moment occurs. On June 9, geologist Rick Hoblitt accompanies an air force general on an aerial survey of Mount Pinatubo. He points out the topography of the volcano, describes the hazards posed by an eruption, and specifically notes the path of previous pyro-clastic flows—right into what is now Clark Air Base. The general turns to his colonel and orders an evacuation. The next day, some 25,000 people are evacuated from within a 19-kilometer radius around Mount Pinatubo. By noon the following day, another 14,500 people leave the air base. Eventually, more than 200,000 people are evacuated.

Two days later, Mount Pinatubo lets loose. Gas-charged magma breaches the surface and blasts ash thousands of meters high. The volcano releases its first pyroclastic flow. It is just the opening salvo, and on June 15, 1991, Mount Pinatubo unleashes what will be one of the largest, most explosive eruptions of the century.

An enormous cloud of gas and ash, hundreds of kilometers wide, rockets 35,000 kilometers skyward (figure 2.3a). Fine ash is driven so high into the atmosphere it is picked up by upper-level winds and spread around the globe. As ash and rock rain down, the sky in nearby villages and across Clark Air Base turns pitch-black. To make matters worse, Typhoon Yunya arrives at about the same time. Hurricane-strength winds sweep the area, and rain turns falling ash into something akin to wet concrete (figure 2.3b). Searing pyroclastic flows and massive avalanches mix with tor-rential rains to create powerful surges of mud and debris that smash through the jungle, flatten towns, and scour away crops.

In the aftermath of the eruption, the countryside surrounding Mount Pinatubo is nearly unrecognizable. Thousands of square kilometers are covered in ash. Volcanic debris and mud blanket mountain slopes and fill valleys with deposits up to 200 meters deep. And at the volcano's summit, removal of magma and rock leaves behind a 2.5-kilometer-wide caldera.

Sadly, more than 350 lives are lost. But given the cataclysmic nature of the eruption and high density of the nearby population, the number of fatalities could have been exponentially higher. Most deaths are caused by roofs collapsing under the weight of rain-saturated ash. Evacuations save an estimated five to twenty thousand lives and property and equipment worth hundreds of millions of dollars. Warnings to military and commercial pilots

Figure 2.3. (a) Mount Pinatubo, 1991 eruption. Courtesy NOAA/NGDC, R. S. Culbreth, US Air Force. (b) Aerial view of pyroclastic flow deposit, Mount Pinatubo, 1991. Courtesy USGS.

save additional lives and prevent millions of dollars in aircraft damage.

What We Learned

The lead-up to and response during the 1991 Mount Pinatubo eruption highlight the value of the Volcano Disaster Assistance Program, which since its inception, has responded to more than seventy-five crises at over forty volcanoes.

The extensive evacuations prior to the eruption showcase both the challenges involved and the success possible. Persuading people that Mount Pinatubo was actually a volcano, and an awakening, dangerous one at that, proved to be one of the many difficulties involved. Sadly, Katia and Maurice Kraft, along with volcanologist Harry Glicken, were killed in a pyroclastic flow while working at Mount Unzen in Japan on June 3, 1991—at the very time other scientists were using their video to educate people about volcanic hazards in the Philippines. In a 1996 retrospective on the Mount Pinatubo eruption, scientists involved suggested that despite their success, they later felt they had been too concerned about overstating the hazards and not concerned enough with speeding preparations for evacuation, adding, "Pinatubo almost overtook us."

From a scientific perspective, the information gained from the 1991 Mount Pinatubo eruption was invaluable. The eruption was found to have involved a mixture of new and old magma. The mixing of fresh magma with preexisting magma is now thought to be a common trigger for eruptions, but the process by which this occurs and its influence on eruptions is not fully understood. Research after the eruption also found that water in the magma might have contributed to the explosiveness of the event. How water influences and drives eruptions is currently an area of intense interest in volcanology.

Based on data from satellite sensors, the 1991 eruption released some seventeen megatons of sulfur dioxide into the atmosphere. The global dispersal of sulfur dioxide aerosols and ash lasted for months and is believed to have caused a 0.5°C decrease in worldwide temperatures. The amount of gas emitted during the eruption was far more than scientists expected given the amount of magma present. They interpreted this to mean that large amounts of gas may have accumulated as bubbles in the magma reservoir rather than dissolved in the magma itself. Once depressurized, the bubbles rose extremely rapidly to the surface and may have caused the eruption to be more explosive (akin to uncorking a champagne bottle). Bubbles, something we take for granted or think of as simple, may play an important and still poorly understood role in volcanic eruptions. Again, this is now an area of active interest in the study of volcanoes.

Another advance made at Pinatubo had to do with the deep, low-frequency or long-period earthquakes detected before the eruption at some distance from the volcano. Relatively distant, long-period earthquakes are now thought to indicate that magma is intruding or breaking through surrounding rock and are specifically looked for as a precursor to eruptions. What happened at Pinatubo also suggests that precursory phenomenon may not be indicative of the size of an eruption and that the run-up to a big event may be as short as a few days. It again became clear that to distinguish anomalous activity from typical background noise, it is essential to have monitoring systems in place as early as possible.

Another important lesson learned at both Mount St. Helens and Mount Pinatubo is that impacts don't end once an eruption ceases. In both cases, long after the eruptions were over, rain events remobilized sediments and debris, causing new and destructive mudflows. After the Mount St. Helens eruption, huge quantities of sediment deposited in rivers had to be removed. Structures have been repeatedly built in an attempt to retain remaining excess sediment that could be washed downstream during periods of high river flow.

The Conundrum of Mount Pinatubo

While the successful response at Mount Pinatubo offers hope to people across the world living near potentially dangerous volcanoes, the eruption itself remains more of an exception than the norm because here is where those pesky wish-we-knews come into play. Not all volcanic eruptions are preceded by signs of unrest, nor are all signs of unrest indicative of impending eruptions. Even when precursors such as earthquakes, ground deformation, or gas emissions occur, they don't necessarily specify when or even if an eruption will happen. It could be days, weeks, or months ahead. Or it may not happen at all.

A buildup of magma and gas can increase the pressure beneath a volcano and drive an eruption. Then again, an earthquake or landslide that removes or releases the overlying pressure can also be the trigger.

Forecasting volcanic eruptions is further complicated by the fact that our record of events is limited to relatively recent history. And because large, catastrophic eruptions occur less often than small to moderate events, we have a better record of minor eruptions—which biases our understanding. While we've come to learn that some volcanoes behave similarly, scientists now understand that their eruptive styles and characteristics vary considerably. Each volcano is an individual; much like people, no two are exactly alike and they change over time. And what about periods of repose or dormancy—what's going on then?

The ultimate goal in volcano science is to amass enough observational data and gain sufficient understanding to accurately forecast eruptions. Though unknowns remain, through research and advances in technology progress is being made. Historical accounts, observations, and the study of geologic deposits from previous eruptions remain critical sources of information, but now added to the pool of investigative tools are space-based satellites and aircraft with cameras or sensors that can take high-resolution images and determine ground deformation, surface temperature, and gravity, and measure volcanic gas emissions. Seismic stations and tiltmeters are used in combination with relatively inexpensive instruments linked to the global positioning system (GPS) to detect ground displacement and deformation. More advanced computers are helping to quickly and precisely analyze seismic data and create improved images of a volcano's interior—in 3-D, or even 4-D. Geochemical analyses of gas, ash, and individual crystals provide insights into how eruptions progress and change over time. Numerical and experimental models are helping us understand a wide range of volcanic processes, such as ash plumes and dispersal, magma movement within the earth, ground deformation, and pyroclastic flows. And infrared cameras are providing a new look at heat signatures, while drones and other remotely operated vehicles are being used in hard-to-access settings and dangerous conditions.

The research highlighted in the following pages is a meager sampling of the studies that are unveiling how volcanoes work. The projects described illustrate what we are learning about volcanoes, but also what experts still wish they knew more about. For ease of understanding and brevity, many of the technical details are omitted. For more on these studies, the science of volcanoes, volcanic hazards, and eruptions, see the list of readings and references at the back of this book.

Volcanoes: The Unknowns

It's in the Plumbing

One of the greatest challenges in studying volcanoes is that their inner workings are hidden from view. Yet the location and amount of magma beneath a volcano and its pathways of flow are critical to understanding how and when eruptions will occur. Imagine how rapid our progress would be if we could see through a volcano's rocky overcoat and watch as magma moved through conduit fractures or emptied from reservoirs. Though we don't yet have Superman's X-ray vision to see into a volcano's interior, modern technology and ingenuity are getting us closer than ever.

Just off the coast of the US Pacific Northwest lies the Cascadia Subduction Zone, where the Juan de Fuca Plate is diving or sinking beneath the North American Plate. Magma generated in the subduction process has created a broad belt of volcanoes to the east, including Mount St. Helens, Mount Rainier, Mount Hood, and Mount Adams. Of the volcanoes in the Cascade Range, Mount St. Helens remains the most active. It is also the site of an ambitious multiyear, multi-institutional effort to reveal the volcano's deep plumbing and how it fits within the regional volcanic setting, and to improve recognition of earthquake activity caused by magma movement or cooling.

Begun in 2014, the iMUSH (Imaging Magma Under St. Helens) project combines in-depth seismic analyses with various geological, geochemical, and electromagnetic studies. As part of a short-term experiment, thousands of seismic sensors were installed on Mount St. Helens. Twenty-three explosions were then detonated to record in detail how the resulting seismic waves passed through the volcano. Because compressive seismic waves travel faster through solid rock than through melt or partially molten rock, the results enabled researchers to create an image of the mountain's plumbing—like a volcano CAT scan.

For many years, volcanoes were envisioned as having one large magma chamber or reservoir beneath the summit and a single conduit to the surface—like those fun models made in grade school. Previous work at Mount St. Helens had suggested that even at shallow depths (3 to 12 kilometers), the volcano's plumbing is not that simple (nor is it at most volcanoes). During eruptions, magma and gas appear to be supplied by a shallow reservoir beneath the mountain. Magma may move into and out of the reservoir as vertical or horizontal slugs of molten rock. Based on results from iMUSH, Mount St. Helens also appears to have at least one other magma reservoir that is deeper, 20 to 40 kilometers down.

Surprisingly, though, the deeper reservoir isn't situated directly beneath the mountain; it is offset to the southeast. It appears that magma is being formed to the southeast, possibly under Mount Adams, then transported west to one or more reservoirs beneath Mount St. Helens. A cluster of long-period earthquakes just southeast of Mount St. Helens, at a depth of 23 to 43 kilometers, may mark one area of magma movement. Other research supports this idea because deep down, directly below Mount St. Helens, the Earth's interior seems to be too cold to produce molten rock or magma. Other studies suggest a puzzling addition to the story—the magmas that have erupted from Mount St. Helens and Mount Adams appear to be geochemically different and not from the same melt source.

When Mount St. Helens is quiet, scientists with iMUSH also detected low-frequency, long-period earthquakes. These are the same type of earthquakes looked for before and during eruptive events. What do they mean when a volcano is not erupting? As more of the data from iMUSH is analyzed, more insights and probably more questions about Mount St. Helens will arise. The researchers involved also hope the results will help to design arrays of monitoring equipment that can more effectively identify magma-related earthquakes and fluid movement prior to eruptions.

A Volcano under the Sea

Another place where scientists are investigating a volcano's hidden plumbing is deep below the ocean, some 500 kilometers off the coast of Oregon. The focus of their attention is Axial Seamount; a volcano submerged under 1,400 meters of seawater. It has a caldera that is several kilometers wide by 100 meters tall and has erupted more than fifty times within the past sixteen hundred years, most recently in 1998, 2011, and 2015. Axial Seamount is also the site of one of the first-ever cabled underwater observatories, part of the Ocean Observatories Initiative. When it erupted in 2015, scientists were already watching and measuring what was happening. Even better, seven months earlier, they had forecasted that an eruption would occur!

Axial Seamount was under scrutiny well before its 2015 eruption. Using instruments deployed on the seafloor, in submersibles, and with remotely operated and autonomous underwater vehicles, scientists had been monitoring the volcano and studying its every move. Based on the data, researchers had identified an amazingly regular pattern of seafloor deformation and earthquakes that precedes eruptions. In particular, as magma pools within a shallow reservoir underlying Axial Seamount, the undersea volcano inflates. The rate of inflation or seafloor deformation varies, up to about 60 centimeters per year. But when its inflation reaches some 3.5 meters, eruptions occur. Knowing this threshold, researchers monitoring the inflation rate could forecast when Axial Seamount should erupt, which is exactly what they did—seven months before the 2015 event.

Scientists studying Axial Seamount also found that when the underlying magma reservoir was drained by an eruption, the seafloor rapidly deflated and the cycle began again. As of early 2018, Axial Seamount had reinflated nearly 1.5 meters. Scientists estimate that if the rate of inflation (deformation) holds steady, it will reach its threshold and could erupt again in three to four years. But there again is a big unknown: Will the supply of magma and deformation remain steady? What in fact controls the supply and rate of magma influx?

Data has also provided invaluable insight into Axial Seamount's elusive plumbing. Scientists have been able to delineate a shallow reservoir of magma about 14 kilometers long and a kilometer thick underlying the seamount. Its structure appears complex. For example, the southeast portion seems dominated by melt or molten rock, whereas in the northwest it is more of a crystal-rich mush—a magma slushy. Not only are the pipes and reservoirs below a volcano more complicated than once thought, what fills them isn't so simple either.

Another interesting finding at Axial Seamount is that seismic activity increases significantly not only before eruptions but also during low tides. The scientists suspect that during low tides the pressure or load of overlying ocean water is minimized so the faults in the seamount unclamp or release, causing more earthquakes.

Work at Axial Seamount has unveiled some of the processes in a volcano deep beneath the sea. In Hawaii, there's a new volcano growing under the ocean—its name is Lōʻihi. But in this island chain, it's an aboveground volcano that's much more renowned. In 2018 Kīlauea volcano put on a dramatic, destructive, and costly show. Based on its past history, the upsurge in activity was not out of the ordinary, but for most people it came as a gushing surprise. And with today's rapid global access to information and imagery, it became a volcanic spectacle unlike any before. The eruption also supplied experts with an extraordinary wealth of data and highlighted some volcano wish-we-knews.

Kīlauea, Hawaii

On the big island of Hawaii, Kīlauea volcano has long been a boon to residents. Fertile, ash-rich soil surrounds the volcano; eruptions have, over time, created new land; and each year more than 1.8 million people visit Hawaii Volcanoes National Park. But in 2018 the benefits of Kīlauea were overshadowed by its fiery and effusive upheavals.

As early as mid-March, there were signs that something was up—or more precisely, that something was going on below. East of the summit, in a zone of fissures or rifts, the ground had begun to inflate. Scientists now interpret this as an indication that magma was accumulating in the ground below. An overflow of the lava lake at the volcano's summit crater in mid-April may have been another indication of an underlying magma buildup. Clusters of earthquakes detected under Kīlauea then began migrating eastward, toward the Leilani Estates subdivision, suggesting that magma was on the move and headed that way.

As Kīlauea's volcanic activity ramped up, cracks and steaming vents opened in the ground, along roads and in backyards. Lava began to ooze and then pour from a series of gaping new fissures, east of the previously active rift zone. Eventually, twenty-four new vents or fissures, some hundreds of meters long, opened up. From one vent shot a towering fountain of glowing lava. It reached up to 80 meters high and built an enormous spatter cone of hardened lava rock that channelized flow to the ocean (plate 6). Rivers and thick massive cliffs of lava burned and bulldozed the landscape. Lava filled and destroyed a locally beloved small lake and picturesque Kapoho Bay. Where burning vegetation released methane, ethereal blue flames shot from cracks in the ground.

During the eruption, a 6.9 magnitude earthquake violently shook Kīlauea volcano. As a result, the flank or seaward slope slumped about 5 meters. The root cause of the earthquake may have been the intrusion of magma from below. It may also have helped to open up the underground plumbing and allow more magma to flow into or through the system.

Meanwhile, at the volcano's summit, the lava lake lowered and disappeared from sight. On average some seven hundred earthquakes shook the ground each day. From the summit crater came periodic explosions of gas and ash. The volcano's larger and surrounding caldera dramatically subsided, slumped, and widened (figure 2.4), while the floor of the inside crater dropped hundreds of meters. Then on August 4, the upheavals of lava, seismic shaking, and slumping at the summit crater declined rapidly, marking the end of the 2018 eruption. But the changes that had occurred, particularly at Kīlauea's summit, left scientists wide-eyed in amazement. And with a real-time camera at the caldera rim and an army of scientists and instruments monitoring the eruption, the public had been able to follow right along with the action.

For people living and working in the area surrounding Kīlauea, the 2018 eruption was devastating. Hundreds of homes were destroyed. More than a thousand people were evacuated. Roads were ruined or became impassable, and a geothermal power plant was heavily damaged. While these impacts, and the sheer volume of lava blanketing the land (more than 35 square kilometers) and running into the sea, were astonishing, based on the volcano's history, they were not all that unusual.

In 1924 Kīlauea's summit lava lake also drained, the surrounding crater subsided, and for several months small blasts occurred. It took years for lava to refill the crater. In 1955 and 1960 lava also poured from rifts east of the summit, covering the land and creating towering spatter cones. In general, with the exception of relatively minor explosive events, Kīlauea's recent history has been pretty mild-mannered. But based on a surprising geologic discovery made a few years back, we know that Kīlauea isn't always a milquetoast volcano. It has a darker and more violent side.

At the volcano's summit, just hundreds of meters from the Hawaii Volcano Observatory and Jaggar Museum (both damaged in the 2018 eruption), are layers of ash and volcanic debris that tell of a very different Kīlauea. In 1790 an eyewitness reported

Figure 2.4. Kīlauea's summit caldera following the 2018 eruption. Courtesy USGS.

seeing a column of ash rising more than 9,000 meters over the volcano. A massive and explosive eruption had occurred, releasing a pyroclastic surge of scorching gas, ash, rock, and debris that sped across the summit. Footprints preserved in an ash layer tell a tragic tale: At the time, a band of battling Hawaiian warriors was marching across Kīlauea's summit. Some escaped, but as many as a few hundred died in the blast. Based on the geology, Kīlauea's 1790 eruption included a series of blasts that topped off several centuries of intermittent, but highly explosive eruptions.

Geologic evidence suggests that in the last fifty thousand years, Kīlauea has erupted violently at least two dozen times. Explosive episodes seem to occur in clusters, each lasting a few hundred years or so. Though violent, these eruptions are nowhere near as explosive as the blasts at volcanoes like Mount St. Helens or Pinatubo. However, an eruption some eleven hundred years ago may have been particularly powerful. During the event, rock fragments 10 centimeters in size were shot out of the crater and, surprisingly, landed east of the volcano. This means they were blown high enough during the eruption to fly above the prevailing westerly-blowing trade winds and enter the easterly winds of the jet stream. Were Kīlauea's fiery upheavals in 2018 the beginning of a new era of more explosive activity? Scientists don't think so.

Beneath Kīlauea, research has identified three regions where magma is stored and supplied to eruptions. One shallow reservoir sits a kilometer or less below the volcano's summit and feeds its lava lake. Somewhat deeper, 3 to 5 kilometers down, and slightly to the south is another, possibly the primary magma reservoir. Below that and beneath the volcano's east rift zone is a less well-defined area where magma seems to permeate the rock. Volcanologist Don Swanson has been working at Kīlauea for decades. He theorizes that the magma reservoir is divided into several receptacles from which magma can leak or flow prior to, during, and between eruptions. Kīlauea's underground plumbing also includes horizontal and vertical conduits and a deep source of melt 80 to 100 kilometers down. This is thought to be the connection to the actual Hawaiian hotspot and a seemingly endless supply of magma.

In general, it appears that magma ascending beneath the volcano is stored in reservoirs and then may erupt at the summit caldera or be transported laterally to a rift or fissure zone (as in the 2018 eruption). As at Axial Seamount, the reservoirs are probably not simple pools of molten rock, but rather complex bodies that can include magma slush—a mix of warm crystals in molten rock.

So what controls Kīlauea's eruptions? In the mid-2000s researchers documented a surprising change in the supply of magma to the volcano. US Geological Survey volcanologist Michael Poland was excited because their monitoring equipment was sensitive enough to detect the twofold increase in magma being supplied from deep within the Earth to Kīlauea. However, after an uptick in the lava flow, the supply of magma waned, and was followed by an eruptive lull.

Based on observations and analyses, researchers don't think a similar increase in the supply of magma caused Kīlauea's upsurge in 2018. Instead, a backup in the plumbing—a proverbial hairball in the pipes—was likely the root cause. The clog in the volcano's plumbing is thought to have come from cooled and thickened old magma, essentially the source of lava that had been slowly oozing from Kīlauea for years. As magma continued to flow from below, pressure built up in the pipes (as indicated by the observed inflation or ground deformation) until a weak spot in the system gave way. Lava then flowed down into the lower portion of the east rift zone. The outflow of old magma also allowed newer, hotter, more fluid magma to enter and flow through the system, which accounted for the faster flows and more vigorous fountaining of lava.

During the 2018 eruption, lava flowed 40 kilometers underground to the rift zone, where it erupted and kept erupting for weeks. A lava rock delta grew at the coast (plate 7) and increased the size of the island by more than 3.5 square kilometers. Overall, the depth of Kīlauea's summit crater more than tripled and the surrounding caldera more than doubled in size. Scientists now estimate that the volume of lava erupted is about equivalent to the collapse area at the summit caldera.

During the 2018 eruption everyone wanted to know what would happen next and when the earthquakes and lava flows would stop? While it remains difficult to predict when a volcano will erupt, it is equally hard to predict how an eruption will progress and when it will end. It appears that once enough lava erupted from the rift zone and the reservoir beneath Kīlauea was depleted, the eruption quieted down. Over time, magma is expected to refill the reservoir and eventually reenter the crater and create another lava lake. But it could take months or even years for this to happen. In 2019 a small pond made a surprise appearance in Kīlauea's summit crater. Scientists think it is linked to groundwater and are watching closely to see what happens next.

One of the big unknowns in Hawaii and elsewhere is water. What role did it play in Kīlauea's 2018 blasts (if any) and in the volcano's more explosive eruptions in the past.

Many volcanoes are wet. Magma contains water, which is also found atop volcanoes as ice and snow, and below as groundwater. When water mixes with fiery magma (at temperatures up to 1,000°C or more) or comes in contact with blazingly hot rocks, it can power explosive blasts, fracture rocks, propel ash, blocks, or bombs skyward, or even generate a pyroclastic flow. Eruptions or blasts caused by water are called "phreatic." In 2014 an unexpected phreatic eruption killed more than fifty people on Mount

Ontake in Japan. In 2017 a smaller, though just as surprising, phreatic explosion injured ten people on Italy's Mount Etna. Water-driven eruptions or explosions are notoriously hard to forecast because they don't necessarily show the same precursors as magma-driven events, such as sulfur or carbon dioxide emissions, harmonic tremors, or ground deformation. Many questions remain about how water interacts with magma. Why, for instance, does it sometimes react explosively and sometimes not? And how often does it trigger eruptions?

During the 1924 eruption, Kīlauea's lava lake dropped dramatically and the crater floor subsided some 180 meters below the rim. Subsequently, the caldera walls collapsed and created a rocky stopper on the top of the volcano. Days later, blasting steam powered a series of explosions that lasted for more than two weeks. Large blocks and rocks went flying, in some cases a kilometer or more from the crater. It has long been believed that the lowering of the lava lake allowed groundwater to seep in and mix with magma or come into contact with the surrounding hot rocks, thus driving the blasts. But based on what happened in 2018 (evidence suggests wall collapses and rockfalls within the summit crater may have disrupted the underlying magma and caused recurring blasts that entrained ash and rock debris), that theory is now in question—going from a thought-we-knew to maybe-not.

Twenty-four hours a day, seven days a week, scientists from the US Geological Survey are monitoring Kīlauea and tracking its eruptions. During the 2018 eruption more than a dozen scientists, as well as students, were flown in to help keep track of and study the volcano's activity. During both active and quiet periods, scientists closely monitor many variables, including temperature, gas emissions, lava lake levels, seismicity, ground deformation, and the geochemistry of erupting magma. During eruptions, researchers work to forecast exactly where and how fast lava will flow. This can be surprisingly difficult because flowing lava creates new topography, which in turn changes its path. Scientists also help local authorities plan and warn residents of hazards, from noxious gas emissions, to possible landslides, earthquakes, and explosions.

As at Axial Seamount, scientists at Kīlauea look for and study patterns in the volcano's behavior, and before the 2018 eruption had observed a remarkably regular pattern of inflation and deflation. As magma filled Kīlauea's underlying shallow reservoir, the volcano gradually (over weeks to years) inflated or deformed. Throughout this process seismometers recorded small high-frequency earthquakes below the summit. Once an eruption began and magma flowed out of the underlying reservoir, an abrupt deflation occurred. The previous type of seismicity stopped and was replaced with low-frequency or long-period earthquakes, which are thought to reflect adjustments in the rock in response to the exiting magma. Unlike at Axial Seamount, however, researchers at Kīlauea have not been able to identify a specific threshold of inflation after which an eruption is imminent. Don Swanson suggests that when ground tilt is high and stays that way for several weeks or months, the chance of an eruption increases, but it is not inevitable. Here again is one of the big challenges in volcano science. Even when a volcano shows significant signs of unrest, an eruption is not a foregone conclusion. Seismic activity, gas emissions, or ground deformation may decrease without a big blow.

At Kīlauea, long-term forecasts of eruptions, a year or more in advance, remain elusive, but based on the general pattern of inflation-deflation and earthquake activity, scientists have been able to provide accurate short-term warnings (hours to months). Scientists have successfully traveled to the site of probable activity and issued alerts before lava broke out onto the surface. But even then, surprises occur.

Volcanologist Matthew Patrick describes a period in 2014 when a lava flow was approaching the town of Pahoa. The lava unexpectedly poured into a deep fracture in the ground and traveled, largely hidden, beneath the surface for nearly 2 kilometers. The lava eventually resurfaced, continued flowing downslope, and reached the town. Fortunately, it stalled before causing major destruction.

Over the years advances have been made at Kīlauea in many areas of volcanology: the study of lava flows and landforms, eruption processes, seismic activity related to volcanoes, gas emissions, the first observations of lava breaking out underwater and cooling, technology development, and more. Yet even here, there remain many wish-we-knews.

With the 2018 eruption, more than ever, scientists wish they knew what exactly is going on beneath the summit: the stepped subsidence, cracking, and slumping of the caldera had never before been observed at Kīlauea. A larger collapse of the caldera was and is possible, but not expected. Matthew Patrick laments that for reasons of safety, eruptions have to be observed from too far away; he wants a closer look and more precise measurement of eruption rates. And like others, he'd like a firsthand view of what's going on beneath Kīlauea—a big viewing window to see how magma and gas evolve and rise during an eruption. Michael Poland wants to better understand what drives volcanic unrest (magma, gas, or water) and, of course, how to tell when precursor activity will culminate in an eruption—or not. Though one of the scientists working at Kīlauea for the longest, Don Swanson still has a myriad of unanswered questions: What causes wall collapses at the caldera, and how can they be predicted? Will there be another period of more explosive eruptions, and when? And he too wants a better view of the Kīlauea's hidden pipes and chambers. Christina Neal, the scientist in charge at the US Geological Survey's Hawaii Volcano Observatory, would like to know more about changes taking place under the sea at the sides of Hawaii's active volcanoes. She is also striving to learn more about volcanic fog, or vog. Thousands of people are impacted by vog from Kīlauea, and while emissions are measured, it remains a challenge to determine to what extent sulfur dioxide is causing the vog, how its downstream plume behaves, and how it impacts human health.

Volcanoes Grow Up

Scientists now think of volcanoes as having a life history. They are born or created, erupt and grow, are sometimes active, sometimes at rest, and eventually die or become extinct. In Hawaii, all stages of the volcanic life cycle are on show. Below the sea surface, there is even an active toddler, an early-stage volcano that has yet to make an above-water appearance. Lō'ihi sits 975 meters below the ocean surface, 35 kilometers southeast of the big island of Hawaii. It has a summit caldera with three pit craters similar to those previously found on Kīlauea. But there's no need to scramble for new real estate. Lō'ihi grows through undersea eruptions and on average appears to be rising toward the sea surface at a rate of about 3 meters per thousand years. It's estimated that it could become Hawaii's next island in anywhere from fifty to two hundred thousand years. Then again, maybe Lō'ihi will make its appearance sooner than expected in a superfueled magma surprise.

Growing with each new eruption, Kīlauea and the adjacent Mauna Loa are in the fast-growing juvenile stage. Eventually, these volcanoes will move away from the underlying hotspot. As they do, they will cool, become denser, and very slowly sink into the Earth's interior—like the islands further to the west that are more elderly and inactive. As volcanoes age, they are also worn down through erosion and weathering, and sometimes parts break off in a very big way.

Hawaiian volcanoes grow upward at their summits, but they also extend seaward at their flanks or sides through eruptions at rift zones. Over decades and with repeated eruptions, a volcano's seaward flank may become overbuilt and unstable. This instability can result in the slow downward creep or slumping of a slope (as is happening at Kīlauea today) or a sudden and large-scale displacement—a giant landslide (as happened at Anak Krakatau in late 2018 and caused a destructive tsunami). Such flank collapses appear to be part of the aging process in Hawaiian-type volcanoes.

High up on the cliffs of Lanai and Molokai islands, scientists made a surprising discovery: deposits of chaotically mixed coral and lava debris. How did the debris get so high up? Based on the geology: an enormous tsunami, up to 155 meters high, created the deposits. And what could have triggered such a towering tsunami? The answer: the sudden and enormous collapse of a volcano's flank.

Geologic evidence suggests that, over the past five million years, there have been at least seventeen huge flank collapses in the Hawaiian Islands. Evidence of flank collapses and colossal tsunamis are found elsewhere as well, including in the Cape Verde and Canary Islands, Papua New Guinea, and Mauritius. Giant flank collapses can also cause or be triggered by large earthquakes and may open new rift zones. Scientists wish they knew more about flank collapses, how they are triggered, and how often they occur. One thing they are pretty sure about, though, is that even with the recent disgorging of lava at Kīlauea and its seaward growth, a flank collapse is highly unlikely. Here, the slope tends to grow and shift through repeated and smaller-scale slumping.

Another question many people in Hawaii would like answered is when will Kīlauea's neighbor, Mauna Loa—one of the world's largest volcanoes—next erupt? Mauna Loa's last magmatic show was in 1984, and since 2014 the volcano has shown signs of unrest. Ground deformation suggests that beneath the volcano and under a rift zone, magma may be degassing or moving in. Shallow and deep earthquakes have increased periodically. But there have also been times of quiescence. As of yet, there is no distinct pattern or indicator that suggests an eruption is imminent.

A particularly confounding question is whether Mauna Loa will follow previous patterns or act differently this time. There's no way to know. Experts feel confident, however, that if Mauna Loa erupts it will start, as in the past, at the summit and then migrate to one of two main rift zones. Christina Neal notes that with vastly improved monitoring capabilities, when Mauna Loa does erupt, the Hawaiian Volcano Observatory hopes to provide a warning at least a few hours before magma reaches the surface—but she admits there are still uncertainties.

Another question is whether Mauna Loa has an underlying connection to Kīlauea. Research suggests that the two volcanoes produce lava that is chemically different, and they seem to erupt independently of one another, but down deep they may have a common magma source.

The big island of Hawaii will continue to be a hub for volcano activity, monitoring, and research, but it is not the only place where scientists are aggregating their efforts to better understand the world's volcanoes and tackle the related unknowns.

Laki, Iceland, 1783

Little known by most people outside of Iceland, the 1783 Laki eruption was nothing short of a volcanic cataclysm, with impacts that spread across the planet. At the eruption's peak, lava didn't just flow from the ground, it gushed out at up to 6,000 cubic meters per second—that's triple the rate of water passing over Niagara Falls! Glowing, geyserlike lava fountains reached up to 1,400 meters high, and enough sulfur dioxide was spewed into the upper atmosphere to create aerosols that are thought to have cooled the Northern Hemisphere by 1.3°C for up to three years. The resulting fatalities numbered in, at the very least, tens of thousands. Why do so few people know about the Laki eruption or, as it is locally known, the Skaftár Fires? Maybe it's because it happened in a remote location where few people lived, at a time when communications were poor. Volcano monitoring had yet to be born, and it wasn't a classic cone-shaped peak that let loose in one massive blast. In Iceland, though, the 1783 Laki eruption is infamous, and scientists across the world now consider it one of the greatest eruptions in modern times.

Weeks before the Laki eruption, earthquakes reportedly shook the lush green pastures of southern Iceland. Then, on the morning of June 8, 1783, an enormous black cloud was seen looming over the area's sparsely populated foothills. By afternoon, ash swirled in the wind and blotted out the sun. Darkness reigned. A light rain began to fall and mix with the ash, which was highly acidic. Leaves became pockmarked with acid-burned holes. The acidic rain irritated the eyes and skin of people and animals. That night, as the ground again trembled, fissures tore open the land. It was the beginning of what would be a lengthy volcanic siege.

Over an eight-month period, a series of ten eruptions occurred, each seeming to follow the same pattern: first, swarms of earthquakes and the opening of fissures; then interaction with the underlying water table, creating steam-driven explosions; then bigger blasts, fueled by magma, gas, and ash; and finally, lava—tons and tons of lava.

Giant surges of lava raced up to 17 kilometers per day down

river gorges and channels. In total, nearly 15 cubic kilometers of lava issued from at least 130 craters along a 27-kilometer-long linear system of vents or fissures. Laki's towering ash clouds climbed as high as 13 kilometers and released megatons of sulfur dioxide, fluorine, and chlorine into the atmosphere. The layers of ash and falling debris created a huge airfall deposit, twice as large as that produced by the 1980 eruption of Mount St. Helens. Melted glacial ice led to massive floods.

For the most part, villages and towns in the area were spared by channelized lava flows and floods. The hazards that proved lethal, both locally and afar, were not the usual volcanic suspects such as smothering mud, suffocating ash, or searing gas clouds. The killer here was more subtle, though no less deadly.

Laki's lethality began early as rain laced with acidic ash descended on southern Iceland. The volcanic emissions were rich in sulfur and chlorine, but even worse was the fluorine. At high enough concentrations, fluorine is toxic, and some 7,000 square kilometers were blanketed by a poisonous dose. Fish went belly up in local rivers and streams, and in less than a year about 60 percent of Iceland's livestock died. Famine and disease spread; people abandoned their farms. Some ten thousand people in Iceland—about 20 percent of the population—died due to severe malnutrition and disease. Much of the vegetation was destroyed, and it reportedly took up to ten years for some plants to return, if they ever did.

Over the months, as the Laki fissures belched out caustic gas and ash, it was carried high into the atmosphere. Upper atmosphere winds carried the gas and related aerosols east across the North Atlantic Ocean. Europe was plagued by a highly toxic volcanic fog. A strange haze was reported further east as well, in Russia, China, and eventually North America. At times, the sun took on an unusual red, pink, and even blue-green hue. In Europe, the haze destroyed vegetation and crops. People with respiratory or heart ailments suffered the most. Extreme weather, crop failures, and famine spread across the Northern Hemisphere. Thousands of people are thought to have died due to related illness or disease, though at the time the cause was unknown.

In southern Iceland today a soft blanket of greenish tan moss blankets the vast fields of dark lava, low cones, and rocky ridges produced by the 1783 Laki eruption. Wide glacial outwash plains, products of the massive floods, line the rivers and slice through the landscape. With soaring cliffs, plentiful waterfalls, and wide green pastures, the surrounding countryside is truly spectacular. Its sweeping grandeur has been featured in films and television shows, including in the popular *Game of Thrones*. The region now appears so quiet and pastoral those unfamiliar with Laki's once monstrous upheavals might wonder why so few people live there.

Prior to the Laki disaster, nearby Grímsvötn volcano had erupted episodically. Did Grímsvötn's activity play a role in the subsequent 1783 Laki event? This is another aspect of volcanoes researchers wish they knew more about: can an eruption at one volcano trigger an eruption at another nearby? And in Iceland or elsewhere, when will the next great Laki or Laki-like eruption occur?

A Land of Fire and Ice

Iceland is a smorgasbord of volcanoes. Nearly every type of volcano and magma can be found there: explosive, effusive, even submarine volcanoes. A third of Iceland's volcanoes are subglacial—blanketed by ice up to a kilometer thick. It is truly a land of fire and ice. While its many and varied volcanoes present dangers, they also provide a wealth of benefits, including fertile soil, booming tourism and recreation activities, and a geothermal underbelly that supplies clean and plentiful energy. Some 90 percent of the homes in Iceland are powered by geothermal heat.

Iceland's wealth of volcanoes spring from its unusual tectonic setting. The island sits atop one of the world's mysterious hotspots as well as a mid-ocean ridge. The mid-Atlantic ridge has literally split and is pulling Iceland apart along a rift that runs from the southwest to the northeast (figure 2.5).

In total, Iceland hosts some thirty active volcanic systems, along with a hundred or so dormant or extinct volcanoes. Each system typically consists of a central volcano and a series of fis-

Figure 2.5. Underwater view of the rift between tectonic plates slicing through Iceland. Photo E. Prager.

sures. As elsewhere, no two of the country's volcanoes are exactly alike, and they may or may not show precursor activity before an eruption. This is especially worrisome because so many of the nation's fiery beasts are blanketed by glaciers.

Ice, especially a lot of it, atop an active volcano is a recipe for disaster. Not only does a kilometer-thick blanket of ice make monitoring difficult, but its rapid melting during an eruption can result in dangerous mud and debris flows, as well as glacial outbursts—sudden, exceptionally swift and powerful floods. Melting can also provide water that powers explosive eruptions and drives ash clouds high enough to pose a threat to aviation. In 2010 the relatively minor eruption of Iceland's Eyjafjallajökull volcano disrupted air travel for nearly a week, grounded a hundred thousand flights and millions of passengers, and cost the airline industry over a billion dollars.

In the wake of Eyjafjallajökull's 2010 eruption, a program called FUTUREVOLC was initiated, making Iceland a supersite for volcano research. Its most active volcanoes—Grimsvötn, Askja, Hekla, Katla, Eyjafjallajökull, and Bárðarbunga—became the focus of intense study and monitoring. And as it turned out, the timing was excellent, because in August 2014, Bárðarbunga began to stir. The eruption that followed lasted six months, was Iceland's largest in 240 years, and provided an unprecedented opportunity for scientists to study the collapse of a caldera—a poorly understood phenomenon that often accompanies large eruptions.

A caldera is defined as a depression, 1 to 1,000 kilometers in diameter, typically found at a volcano's summit. Calderas are thought to form by collapse after magma empties from a reservoir, leaving the overlying rocky peak without support. Scientists have long wanted to know whether eruptions cause calderas to collapse or if a caldera collapse can trigger eruptions. And when a caldera collapses, is it gradual or does it go all at once? Large collapses happen infrequently, so very few have been observed in modern history—and that's where the 2014 eruption of Bárðarbunga comes in.

Located in a remote part of central Iceland, Bárðarbunga lays under Vatnajökull, one of country's largest glaciers. The volcano's roughly oval-shaped caldera is 11 kilometers long by 8 kilometers wide and filled with hundreds of meters of ice. And when the 2014 eruption began, scientists were there, monitoring earthquakes, ground deformation, and more.

Based on the data collected, the eruption began as magma rose from a reservoir some 12 kilometers below the volcano. At first, earthquakes marking the underground movement of magma clustered below the southeast corner of the caldera. After a few hours, they began to migrate further to the southeast. Magma was moving laterally underground. By tracking the earthquakes, researchers followed the molten rock as it traveled some 7 kilometers to the southeast and then took a sharp turn to the northeast. Over the next two weeks scientists continued to track the magma's subterranean travels based on ground deformation and the migrating pattern of earthquakes. As the eruption progressed, the magma flowed another 41 kilometers to the northeast, at a depth of 6 to 10 kilometers. Eventually magma broke through to the surface and poured out as lava.

From start to finish, scientists carefully monitored Bárðarbunga's summit caldera. Data indicated that after some 12 to 20 percent of the magma had flowed out of the reservoir below the volcano, the caldera began to subside. Then, as the reservoir continued to drain, the caldera collapsed at a steadily declining rate, helping to drive and regulate the flow of magma. Overall, the caldera collapsed some 65 meters, creating a 110-square-kilometer ice-filled bowl at Bárðarbunga's summit.

Scientists were thus able to show that the collapse of the caldera was gradual and began after the eruption. In other words, the eruption triggered the collapse and not the other way around. Researchers suspect that if the magma had not moved laterally underground, but instead had risen directly to the volcano's ice-covered summit, the eruption would have been more explosive.

Was Bárðarbunga's behavior illustrative of what happens at other volcanoes during caldera collapse? When Mount Pinatubo blew in 1991, the caldera collapsed in a stepped or punctuated manner rather than gradually over time. In 2018 Kīlauea's caldera subsided, slumped, and grew in steps, but a larger collapse didn't happen. What about the world's supervolcanoes, those that have the potential to disgorge cataclysmic amounts of material—more than 1,000 cubic kilometers? Do they behave like Bárðarbunga, like Pinatubo, or in a manner as yet unknown?

Experts in the past have suggested that Earth's mega- or supervolcanoes erupt every hundred thousand years or so. That estimate has recently been revised to something like once every seventeen thousand years. Still, no one is expecting another mega-eruption anytime soon. But far away from Iceland is a place where researchers are concentrating their efforts to learn more about caldera-forming eruptions and big volcanoes—in this case, a supervolcano.

Yellowstone National Park, Wyoming

Yellowstone is one of the most volcanically active regions on the planet. Hundreds to thousands of earthquakes shake the area each year, and the ground moves up and down, annually inflating and deflating about 7 centimeters. The area hosts extraordinarily high heat flow and the world's largest concentration of boiling mud pots, towering geysers, and steaming hot springs. But today's action is nothing compared to that of the past, because Yellowstone is one of the world's so-called supervolcanoes.

In the past two million years or so, Yellowstone has had three mega-eruptions: 2.1 million years ago, 1.2 million years ago, and the most recent, 630,000 years ago. Together, these eruptions may have produced enough ash and lava to fill the Grand Canyon. Numerous smaller eruptions have occurred as well, including one about 70,000 years ago.

What would happen if Yellowstone were to erupt today as it did 630,000 years ago? It would be a cataclysmic disaster of a magnitude unseen in modern history. Volcanic debris would be hurled hundreds of kilometers and reach as far as Louisiana, Iowa, and California. The amount of ash and rock blasted from the caldera could be more than twenty-five hundred times that of the 1980 Mount St. Helens eruption. Ash would be spread across the western US for tens of thousands of kilometers. Sulfur dioxide shot into the upper atmosphere would cause cooling that could last up to a decade. Global climate and weather patterns would be changed for years. Rivers and sewer systems would be inundated and clogged by debris. Infrastructure integrity would be tested by earthquakes and fallout, with transportation, agriculture, power, and other systems disrupted on a massive scale. Human health would be compromised by diminished air quality, along with shortages of potable water and, over the longer term, food.

But don't hit the panic button. Scientists studying and monitoring Yellowstone say there is no sign that an eruption is imminent and that if one were to occur, there's little chance it would be another colossal supereruption. Researchers are confident that for a significant eruption to take place, a lot of magma would need to be mobilized. And if that were to happen, ongoing monitoring efforts would surely detect the associated changes in earthquake activity, gas emissions, and ground deformation well in advance. That's not to say there aren't still big questions about Yellowstone, its underlying hotspot, and what drives its upheavals.

Through years of research and monitoring, Yellowstone's underground plumbing is coming more sharply into view. Seismic data from local and distant earthquakes have revealed the presence of dual magma reservoirs, with a combined volume of more than 10,000 cubic kilometers. That's large enough to hold 182 Mount Everests, and if they were full of molten rock, it would be a heck of a lot of eruptable magma. But that doesn't appear to be the case—at least for now.

One of Yellowstone's magma reservoirs sits in the Earth's upper crust about 12 kilometers below the surface. It is estimated to be about 90 kilometers long, 5 to 17 kilometers deep, and tilted up toward the southeast (figure 2.6). This reservoir is not filled with molten rock; it appears to be more like a rocky sponge, with pockets of fluid or melted rock. The amount of molten material is important, because eruptions tend to occur only when the proportion of melt exceeds 50 percent. In this upper magma chamber, researchers have detected between 5 and 32 percent melt. But there's a pesky wish-we-knew here: it remains a challenge to distinguish solid from molten rock. We can't see directly under the volcano, so our measurements are blind, relying on indirect methods. Researchers aren't sure if the resolution of their measurements is good enough to detect all of the molten rock present. Could there be more there than we think? And is it possible, even if the reservoir as a whole poses little risk, that individual or small pockets of molten material could erupt?

Yellowstone's second magma reservoir is more than four times the size of the shallow or upper reservoir and appears to lie between about 20 and 50 kilometers down. The chemistry of the magma here appears to differ from that in the reservoir above, and the melt is estimated at only about 2 percent. In other words (despite reports from some questionable news sources), it is not a ginormous bowl of molten rock waiting to erupt. Relatively new data has added to the story and outlined a hot plume of magma *below* Yellowstone's lower magma chamber. It appears to be cylindrical, 350 kilometers in diameter, tilted to the northeast, and rising from the outer core-mantle boundary, some 3,000 kilometers down—almost halfway to the center of the Earth.

So how does Yellowstone's plumbing work? It's not crystal

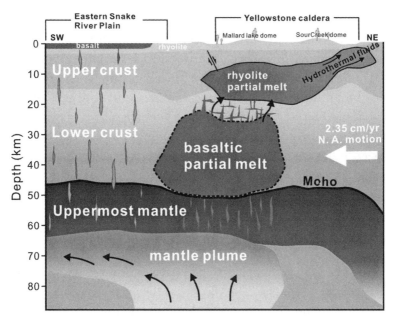

Figure 2.6. Using seismic waves, scientists are beginning to reveal Yellowstone's hidden plumbing and the complexity of what lays below. Courtesy Hsin-Hua Huang/Academia Sinica.

clear, but maybe something like this: A hot plume of magma ascends from deep within the Earth to below the lower magma chamber. Slugs of molten rock move from the plume up through vertical fractures into the large lower reservoir. From there it can travel either upward or laterally, eventually reaching the shallow reservoir. But what controls the magma supply, and what triggers eruptions? Those answers remain in the category of wish-we-knew.

One volcanic unknown being investigated at Yellowstone is how long it takes for the underground system to become primed for an eruption. Based on the geochemistry of crystals found in a fossilized ash deposit from Yellowstone's eruption 630,000 years ago, some scientists estimate it could take mere decades for the magma below to mobilize before an eruption. Other research, however, suggests it could take much longer—thousands to hundreds of thousands of years. It's another magma mystery scientists continue to tackle.

Volcanologist Jacob Lowenstern and his colleagues want to confirm what we think we know about Yellowstone's subterranean plumbing. Simply put, they want to drill. However, drilling deep into the Earth is expensive, time-consuming, and technologically difficult. Yet it is the one way for scientists to get that oh-so-desired direct view of what's really happening underground. When drilling has been done, the findings have included some science surprises. At Katla volcano in Iceland, drilling unexpectedly discovered molten rock in an area where it was supposed to be solid. In Long Valley, California, where experts thought there would be magma or hot rocks, deep drilling revealed surprisingly cool temperatures. What would drilling in Yellowstone unveil?

As for Yellowstone's famous bubbling mud pools and towering geysers, they appear to be fueled by water from rain and melted snow that percolates deep into the Earth's crust, is heated by magma and hot rocks, and then returns to the surface. The round trip may take hundreds to thousands of years. A project is now under way to map the pathways water takes—the plumbing through which it circulates beneath Yellowstone and how it drives the region's hot and steamy activity.

Yellowstone illustrates another of the great challenges in volcano science: forecasting eruptions in volcanoes that have long been dormant or in repose. As Don Swanson puts it, there are no stages in quiet. A volcano may be quiet for a year, a thousand years, or much longer. Without an eruption in modern times and with much of the geologic data buried, eroded, or simply inaccessible, it is especially difficult to forecast when another eruption will occur or exactly what it will be like. But again, no worries, Yellowstone remains a sleeping giant, and scientists don't expect it to awaken anytime soon.

Laguna del Maule, Chile

Which of the world's volcanoes is potentially the most dangerous, the most deadly? Which volcanoes worry scientists the most? Some that make the list: Merapi and Sinaburg in Indonesia, Katla and Heckla in Iceland, Cotopaxi in Ecuador, Vesuvius and Campi Flegrie in Italy, Popocatepetl in Mexico, Nyirangongo in the Congo, and Mount Rainier and Mount St, Helens in the Cascades. But in recent years, a little-known volcanic area in Chile has drawn the attention of the world's volcanologists. Since 2007, at a location known as Laguna del Maule, the Earth's crust has been inflating at an astonishingly rapid rate, more than 20 centimeters per year. With that much uplift, scientists think there must be a large and potentially growing intrusion of magma underground.

Located in central Chile, near the Argentina border, Laguna del Maule is composed of a cluster of small volcanoes, lava domes, ash deposits, and a large, lake-filled caldera. Geologist Brad Singer has been leading efforts to better understand and monitor the region. He and his team have discovered that within the past million years or so, there have been several eruptions a hundred times larger than the 1980 Mount St. Helens event. Based on their data, there might be as much as 100 cubic kilometers of magma just 5 to 15 kilometers below the surface. But again, it is not a simple picture underground. The magma chambers and conduits here may contain molten rock, a crystalline mush (or magma slushy), and sections of cooling and hardening rock.

A single reservoir appears to exist beneath Laguna del Maule and has sometimes fueled eruptions, while at other times magma has risen and cooled more slowly, producing giant bodies of rock essentially frozen in place. What determines which will happen next is another wish-we-knew. But given the potential for an eruption of enormous proportions, the scientists are working hard to better understand Laguna del Maule's plumbing and determine if it is priming for another big blast. Singer claims he and his many colleagues will "throw everything, including the kitchen sink, at the problem—geology, geochemistry, geochronology and geophysics—to help measure, and then model, what's going on."

Our Volcanic Future

On planet Earth volcanoes create land and fertile soil. In some regions they provide clean, abundant energy and fuel lucrative tourism and recreation industries. Yet volcanoes also pose serious and life-threatening dangers. Eruptions are going to happen. Many will occur gently and without much fanfare, while others will be explosive, deadly, and possibly on a scale we have never seen. And with escalating population growth, an increasing number of people and sectors of society will be impacted. But as Marta Calvache, a volcanologist and director of Colombia's Geologic Service, explains: volcanoes aren't the problem, it's the people; we have to learn about volcanoes, because we are living with volcanoes.

As our understanding of volcanoes grows, some of today's scientific unknowns will become knowns. New technology will bring fresh insights, and additional questions are sure to arise. But we don't need to know everything about volcanoes to prepare for eruptions and do our utmost to protect lives and property. We already know that without ongoing monitoring volcanic eruptions cannot be forecasted nor accurate alerts provided. Yet only a small percentage of the world's volcanoes are even partially monitored. And many, if not most, communities at risk have no early-warning system or coordinated disaster plan; residents may not even be aware of the potential hazards posed by a nearby volcano.

The Volcano Disaster Assistance Program is and will be critical in efforts to prepare for and respond to volcanic unrest. But it is not enough. Humans by nature focus on the short-term. And when there has not been a major volcanic disaster in some time, it is easy to forget the past and ignore the realities of the future. While it is imperative to continue to learn about volcanoes, it is also critical that we invest in more monitoring and preparation. Because the unknown here is not whether a volcano will erupt. It's when and where it will happen next.

3

Earthquakes and Tsunamis

We can't assume anymore that there's a subduction zone that can't produce these very large subduction-zone earthquakes and tsunamis.
—**Jeanne Hardebeck**, seismologist, US Geological Survey

INDIAN OCEAN, OFF THE NORTHWEST COAST OF SUMATRA, 2004. For a hundred years, a battle raged. The combatants were two tectonic plates converging in the subduction zone beneath the Sunda deep-sea trench in the Indian Ocean. The Indian Plate was trying to force its way under and past the overlying Burma Micro-Plate (part of the Sunda or Eurasian Plate). But it was a no-go; the plates were stuck, locked together at their roughened and sticky edges. As the stalemate lingered on, a tremendous amount of energy built up within the interface, or fault, between the plates. When the deadlock finally broke, it was abrupt and violent.

On December 26, 2004, some 30 kilometers below the sea-floor, the Earth gave way in a planet-thumping tear. The fault between the two tectonic plates began to rupture. At first it tore slowly, but then accelerated, ripping apart at 2.5 kilometers per second. Within minutes, the rupture stretched some 1,200 to 1,500 kilometers, roughly the length of California, and may have broken through to the seafloor. The result: a 9.1 magnitude mega-quake and the upward thrusting of the Burma Micro-Plate.

As the Burma Micro-Plate went up, the seafloor rose several meters and with it went the overlying ocean, which was thousands of meters deep. All along the rupture a massive amount of seawater was shoved upward. The energy in the uplift was quickly transferred into wave motion. The entire water column, thousands of meters deep, was set in motion. Long, low waves (50 centimeters high) raced outward at the speed of a jet airplane—some 800 kilometers per hour. The greatest amount of energy propagated perpendicular to the ruptured fault. In some areas, undersea ridges and valleys steered and amplified the waves' energy. As the waves traveled through the open ocean, little energy was lost. At the surface in the deep sea, the colossal tsunami was hardly noticeable.

Fifteen to twenty minutes after the earthquake, the long, low waves heading east approached the northwest coast of Sumatra. As the depth decreased and the sea grew shallow, friction from the seafloor caused the base of the waves to slow, while the bulk of the overlying water rushed forward unperturbed. This caused the waves to steepen and grow in height. Where coral reefs lined the shore, the waves encountered even greater friction due to the complex and shallow underwater topography. Some waves broke and released a portion of their energy. Other waves continued to grow and gather, continuing toward shore at about 30 to 40 kilometers per hour. In some areas, the shape of the coast acted like a funnel, and the waves grew even higher.

Before the tsunami struck, water was pulled offshore into a growing wall or surge of water. Fish left stranded by the receding sea flapped helplessly on the sand. People ran out to collect the fish or to observe the strange behavior of the ocean. Moments later, the first in a series of enormous waves or surges hit. Where mangroves lined the coast, the tsunami's energy was dampened and its flow impeded. But low-lying and more open areas were especially vulnerable to the oncoming stampede of seawater. In Banda Aceh, the capital of Aceh province in Sumatra, seawater surged inland as far as 2 kilometers. Where amplified by the shape

Figure 3.1. The aftermath of the 2004 megaquake and tsunami that destroyed Banda Aceh, Sumatra, Indonesia. Courtesy US Navy/Photographer's Mate 3rd Class Tyler J. Clements.

of the coast, the incoming waves reached unthinkable heights, greater than 30 meters. Then, just as the ocean had raced ashore, it rushed back, creating strong currents that dragged anything and everything seaward.

The damage and destruction were nearly unimaginable (figure 3.1). Buildings, bridges, and roads that had withstood the megaquake were wiped away by the tsunami. Homes and hotels collapsed or were simply swept away. A third of the residents of Banda Aceh died or went missing. Meanwhile, out in the Indian Ocean, the waves were still traveling, taking aim elsewhere.

One to two hours after the earthquake, the tsunami struck and wreaked havoc in Thailand. Two to three hours after the earthquake, it hit Sri Lanka. The waves continued on their path of destruction to the Maldives, Mauritius, Africa's east coast, and beyond—literally traveling around the world. In all, some eighteen countries were impacted. More than 230,000 people were killed, half a million injured, and 1.7 million left homeless. The devastation across the Indian Ocean was unprecedented in modern history. Estimated losses were on the order of $12 billion.

The megaquake and tsunami of December 26, 2004, were another global wake-up call. The tragedy brought earthquake and tsunami science to the forefront. At the time, there was no way to detect or track tsunamis in the Indian Ocean, nor were there means to issue or communicate warnings. Even in areas that could have had hours of warning, the deadly waves came as a surprise. Investment in tsunami research, education, and an Indian Ocean warning system became a priority. With the tragedy came hard lessons, and once again some aspects of the event surprised even the experts.

Before the 2004 megaquake, most scientists did not think the northern section of the Sunda Subduction Zone off the west coast of Sumatra could produce such an enormous earthquake. Accepted thinking at the time was that the sinking Indian Plate was too old and moving too slowly to build up sufficient strain. Experts were also surprised by the height of the tsunami generated and by the damage it caused. The inundation and destruction were far greater than predicted by models. The manner in which underwater valleys and mountains (seafloor bathymetry) had focused the tsunami's energy along specific paths, essentially beaming it through the sea, was also unexpected. Channeled by the mid-ocean ridges, the wave energy traveled much farther than it would have otherwise (plate 8).

After the event, teams of researchers fanned out to collect data and survey the impacted areas. They found that the fault rupture began slowly and then progressed to a weak area that let loose more violently. Measurements of ground deformation suggested that the fault continued to slip even after the seismic activity stopped. Plus, the ruptured fault lay below and affected very deep water. These factors are thought to have contributed to the enormous size of the resulting tsunami. It was also discovered that it wasn't just waves striking the shore that did so much damage but also the powerful seaward flowing currents that were generated.

No event of comparable magnitude had occurred in the region in historical times. But afterward, scientists began looking for geologic evidence of earlier tsunamis. And they found it. We now know that in the last 300 years there have been some sixty-nine tsunamis in the Indian Ocean region.

In Aceh, Indonesia, researchers made a particularly spectacular and unexpected discovery: a coastal cave whose sand deposits record thousands of years of tsunami history. The record represents a period between twenty-nine hundred and seventy-four hundred years ago during which there were at least eleven tsunamis in the area. The data suggest that the recurrence interval or time between tsunamis was on the order of hundreds of years, but was also highly variable, with possibly longer quiet periods in which none occur. The bottom line: we cannot predict when tsunamis will strike in the Indian Ocean, but we now know they have happened in the past and will occur again in the future.

By coincidence and for the first time, data from a satellite passing overhead at the time of the tsunami gave scientists actual measurements of the waves as they moved through the open ocean. When combined with large-scale laboratory experiments, this and other data collected following the 2004 Sumatra earthquake and tsunami allowed researchers to greatly refine numerical models used to simulate and forecast the generation, travel, and impacts of tsunamis.

More than ever before, the events of December 2004 illustrated that subtle complexities in how and where an earthquake occurs play a huge role in the size and shape of the tsunami generated. Factors such as location, seafloor bathymetry, and the composition and nature of the shoreline were shown to influence the impacts of a tsunami. Scientists also began to suspect that other subduction zones once thought incapable of producing a megaquake and large tsunami might be vulnerable as well. In 2011 another catastrophic disaster occurred and furthered some of the lessons learned in 2004, while also unveiling some new surprises.

EAST OF TOHOKU, JAPAN, 2011. On March 9, 2011, a 7.2 magnitude earthquake rocked Japan. Within two days, three more temblors struck, all with a magnitude of 6.0 or greater. It was nothing unusual as Japan is a country well versed in and prepared for earthquakes. Japan lies to the west of no fewer than five deep-sea trenches indicative of subduction zones: the Kuril Trench, the Japan Trench, the Izu-Bonin Trench, the Nankai Trough, and the Ryukyu Trench. In the northeast, the massive Pacific Plate moves westward and is diving beneath the eastward-moving Eurasian Plate. In the southwest, the Philippine Plate is sinking or being subducted under the Eurasian plate. As the tectonic plates in these junctions jostle, stick, and slip, earthquakes happen. In Japan, they happen a lot. It is a way of life; shake-resistant structures have been built, people are educated about what to do in the event of an earthquake, and an early-warning system is in place.

But on March 11, 2011, something unexpected happened off the coast of the Tohoku region. Just a few days after the 7.2 magnitude earthquake, some 20 to 30 kilometers below the seafloor under thousands of meters of ocean, the fault between the Pacific and Eurasian Plates gave way. At first it was a deep, quick, and violent tear, sending seismic waves radiating outward. Early estimates of the earthquake's magnitude were low, based on the initial rupture. But the rupture didn't stop as expected; instead it propagated along the fault between the two plates, moving slowly up through the crust and water-filled sediments toward the surface. As a result, the Eurasian Plate, once tightly wedged-in over the Pacific Plate, slipped and pushed the steeply sloping seafloor eastward. And with it went the overlying 7 kilometers of ocean.

As in the 2004 Sumatra quake, the displacement of so much seawater triggered a colossal tsunami. Energy in the form of long, low waves raced across the open ocean. Fifteen to twenty minutes after the earthquake, the tsunami struck the east coast of Japan. Where the coastline narrowed into a bay, the waves were amplified, creating towering walls of water as high as 39 meters. Rushing kilometers inland, the tsunami wreaked havoc, especially in low-lying areas.

The images beamed across the world were shocking, with cars and boats toppled over like toys (plate 9). Water surged over and across seawalls. Homes and multistory buildings were sheared apart and swept away. Debris-laden currents raced across the land. At least sixteen thousand people were killed, and thousands remain missing. Some 230,000 homes were lost and an estimated five million tons of debris washed offshore. The total cost has been estimated at a mind-blowing $235 billion, including the catastrophic and infamous failure at the Fukushima Daiichi nuclear complex. It was the largest earthquake in 140 years in Japan and one of the most costly disasters in modern times.

The 2011 Tohoku earthquake, now recognized as a 9.0 magnitude megaquake, the tsunami it generated, and the destruction they caused added up to another wake-up call for experts and laymen alike. Based on the segmented or broken-up nature of the fault in the subduction zone and its history of being locked and sometimes creeping (slow slip), the majority of scientists predicted that at most a 7.5 magnitude earthquake could occur. So that's what was planned for. But a handful of scientists had suggested that a larger quake was possible. Based on ground deformation, geologist Yasutaka Ikeda warned that a 9.0 magnitude earthquake could hit the Tohoku region. Furthermore, while there were no modern records of a giant quake in the region, geologic deposits indicated that a great tsunami had occurred in Sendai in 869 AD; this led paleontologist Koji Minoura and his colleagues to also suggest a megaquake was possible. At the time, some experts were even discussing how to integrate prehistoric events into modern risk assessments. But many people, including experts, still did not believe such a large earthquake was possible in the area.

Given Japan's history of earthquakes and investment in science, an array of seismic monitoring instruments was present during the 2011 event. Data indicate that the fault rupture began at depth (20 to 30 kilometers down) and slowed, but then tore all the way up to the surface. This was a surprise; most scientists didn't expect a large slip or rupture in the shallow section of the fault. And as Woods Hole Oceanographic Institution geophysicist Jeff McGuire explains, the fault motion actually got stronger as it got to the surface. That's why the tsunami was so big.

In total, five fault segments ruptured, and the two tectonic plates slipped past one another some 50 meters or more. A giant slab of seafloor, the size of Connecticut, jumped 10 meters vertically. The earthquake thus grew to a 9.0 magnitude monster and generated a disastrously large tsunami. Researchers also found significant spatial variation in the slip and determined that the event may have caused slumping or a landslide on the seafloor.

How the fault ruptured during the Tohoku event highlighted our lack of understanding about how faults tear and behave at depth. And it again illustrated that tsunami size is not simply a function of earthquake magnitude, but also the depth, length, and speed of the fault rupture. After the disaster, scientists began digging through the record of earthquakes leading up to the tragic event. Through intense investigation and with the help of precise measurements made by instruments on the seafloor, they discovered several slow-slip or silent earthquakes (more on this later in the chapter) just to the north before March 11. Some scientists think these events may have provided the last bit of strain needed to start a rupture or tear on the fault—the proverbial straw that broke the camel's back. This is indeed one of the great unknowns in earthquakes: how do they actually start, or for that matter stop?

Another takeaway from the Tohoku event is that even a highly segmented fault within a subduction zone with variable slip behavior can produce a catastrophic megaquake. This, combined with the 2004 Sumatra disaster, made scientists again consider which, if not all, of the world's subduction zones are capable of producing megaquakes and destructive tsunamis.

Using the Japanese deep-sea drilling ship *Chikyū* another team of scientists drilled into the fault that ruptured in the 2011 disaster. They discovered that in the area that broke, the fault between the Pacific and Eurasian Plates is unusually thin, less than 5 meters thick, and lined with slippery clay. Both of these factors may have weakened the fault and contributed to its surprising behavior during the rupture. Another theory is that the wedge of sediments within the trench overlying the subduction zone may have amplified the energy released. These findings raise another question: Is this unique to this one subduction zone, or could others behave similarly?

The disasters of 2004 and 2011 were watershed moments in the science of earthquakes and tsunamis. The lessons learned not only changed how we prepare for and respond to such events, but also underscored how much we still have to learn. Faults are no longer thought to produce characteristic same-size earthquakes that occur at regular intervals. Scientists have had to rethink and revise how models simulate earthquakes and tsunamis, and reconsider how and where earthquakes can grow to catastrophic proportions.

These events and others also remind us that modern records are all too often too short to identify the scale of the potential dangers posed. Today, teams of researchers across the world are looking back in time to increase our knowledge base, continuing their efforts to better understand earthquakes and tsunamis, and to help people at risk prepare. As usual, with greater understanding come new questions, and even as we advance, there remain many elusive unknowns and wish-we-knews.

Earthquakes: The Known

Thousands of earthquakes occur every year. Most are too small to feel. But each year there are also more than a hundred earthquakes of magnitude 6.0 or greater. And every several years, one or two true monster shakers strike; these are earthquakes of magnitude 8.0 or greater. They literally ring the planet and can cause devastating loss of life and property. Unfortunately, progress in our understanding of earthquakes is often coincident with destructive events. One relatively early event resulted in a seismic leap in the study and understanding of earthquakes.

San Francisco, 1906

On April 18, 1906, San Francisco residents received a truly unwelcome and unexpectedly violent early-morning wake-up call. Some heard a loud roar, while others felt a strong jolt or were literally shaken from their beds. The earthquake reportedly lasted for what seemed a terrifyingly long time. Buildings toppled. Gas mains and pipes were severed. A citywide inferno ensued as ruptured water mains made it impossible to fight fires fueled by unseasonably warm temperatures and high winds. Thousands of people were killed and many more left homeless. It was one of the nation's most devastating disasters and sparked an intense investigation into the causal earthquake.

Scientists from all across California responded to the call for help in investigating the great 1906 earthquake. They surveyed, collected observations, analyzed seismic records, and literally dug into the ground. To synthesize the results of their efforts, California's governor appointed a State Earthquake Investigation Commission. Its report, now known as the Lawson Report, was a windfall for earthquake science. It detailed the damage wrought by the San Francisco earthquake, catalogued seismic data from instruments around the world, and described a fundamental concept in what we think, still today, happens during an earthquake.

The Lawson Report included data showing that the earthquake was the result of a rupture 435 kilometers long on a newly named fault, the San Andreas Rift. In an instant, straight roads, streams, tunnels, bridges, and fences that crossed the fault had been offset 3 meters or more. The data demonstrated that earthquake damage depended both on the design and construction of buildings and on local geology. Some of the strongest shaking was shown to have occurred in soft soils, especially areas of fill.

The availability of pre-quake survey data made it possible for scientists to deduce that prior to the event the land had deformed and then, during it, had slipped or become offset. Geologist Harry Fielding Reid traced the deformation and offset and came up with an explanation for what happened during the earthquake. He theorized that some distant source of stress had caused a gradual buildup of strain in the area and that this caused the ground to deform. During the earthquake, the Earth's crust had literally snapped and then rebounded elastically—imagine a rubber band being stretched until it breaks. At the time, Reid could not explain what caused the strain in the San Francisco region, but

Plate 1. From a spacecraft orbiting more than 320 kilometers above Earth, the atmosphere appears as a thin blue band encircling the planet. Courtesy NASA/JPL/UCSD/JSC.

Plate 2. Antarctic summer meltwater runoff and Nansen Ice Shelf waterfall. Courtesy Won Sang Lee, Korea Polar Research Institute.

Plate 3. Traffic jam of icebergs in the Kangia Icefjord, summer 2017. Note hikers in foreground for scale. Photo E. Prager.

Plate 4. The Great Ocean Conveyor Belt with overturning flow of deep cold waters (blue) and warm surface waters (red). Courtesy NASA/JPL.

Plate 5. Coral bleaching in Cayman Brac. © Stephen Frink.

Plate 6. Lava pours out of a 17-meter-high spatter cone built around fissure eight during Kīlauea's 2018 eruption. Courtesy USGS.

Plate 7. Lava enters the sea from fissure eruption along a kilometer-long stretch of the coast in the 2018 Kīlauea eruption. Courtesy USGS.

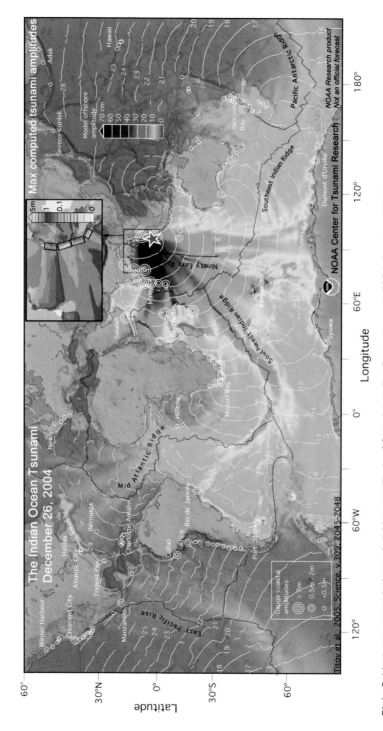

Plate 8. Maximum computed wave height or amplitude of Sumatra, Indian Ocean tsunami, 2004. Courtesy NOAA.

Plate 9. Tsunami wave approaching Miyako City from the Heigawa estuary in Iwate Prefecture during the 2011 Tohoku disaster. HD/Reuters.

he estimated it would take another two hundred years for it to build up and cause another earthquake of similar magnitude. His "elastic rebound theory" remains fundamental to how we think the Earth's crust behaves during an earthquake. Today, however, we know the mysterious source of Reid's strain—plate tectonics.

Though a wish-we-knew in 1906, it is now well understood that the San Andreas Fault is the boundary between two moving tectonic plates, the juncture where the Pacific Plate moves northwest relative to the southeast motion of the North American Plate. Because the edges of the plates are irregular and rough, friction causes them to get stuck or locked, and so, over time, strain builds up. When a threshold is reached, the San Andreas, like other faults, ruptures, causing an earthquake and a slip or offset. This type of fault, where two tectonic plates or blocks of crust are moving horizontally in opposite directions past one another, is known as a transform fault.

When the global distribution of earthquakes is plotted on a map of the Earth's surface, it produces a clear and at one time startling picture. The majority of the world's earthquakes line up along or near the boundaries of the tectonic plates. In fact, this was one of the critical clues that led to our understanding of plate tectonics. Where the plates converge, move apart, or slip horizontally past one another, friction acts to impede their movement. Sometimes they get stuck or locked, and then, as the plates continue trying to move, the surrounding crust or ground deforms. The elements that contribute to earthquakes, as geophysicist Ross Stein explains, are the steady motion of the Earth's rigid tectonic plates; the rocks that compose the plates, which can act rubbery or elastic; and friction, which causes stickiness. When the strain is too much, friction is overcome and the Earth's crust breaks along weak points, or faults, typically the boundary or interface between two plates. These faults can be vertical, sloped, hidden at depth, and segmented.

We now also know that rocks behave differently with depth in the Earth. In shallow depths, the Earth's crust is brittle and can fracture or break under strain. Deeper down, as temperature and pressure rise, the Earth appears more ductile and rocks move more fluidly without the buildup of strain and breaking. The Earth's crust is also not homogenous but varies in composition and structure, and this too changes how it behaves under strain. And we know that earthquakes are not one-shot deals but happen repeatedly over time as strain builds up, is released, and then builds up again. But as scientists have come to learn, this is not a simple or regularly repeating process.

At the seafloor where tectonic plates diverge or move apart, new ocean crust is formed. Here the Earth's crust is relatively young, hot, and weak, so not a lot of strain builds up and earthquakes tend to be relatively small. In subduction zones, where plates converge and one sinks under the other, the crust tends to be older and cold, thus making it stronger and more able to absorb elastic strain. This is one reason larger earthquakes tend to occur in subduction zones.

During an earthquake the accumulated energy is released into the surrounding rock as vibrations or seismic waves. The strength or size of an earthquake is dependent on the amount of energy released, which is a function of the size of the fault rupture and offset. Typically, an earthquake's aftershocks (the relatively smaller quakes that occur after the main shock) outline the rupture of a fault. In recent decades, scientists have discovered that strain can also be released by aseismic creep, or slow slip, that doesn't produce earthquakes (more later in the chapter).

Earthquakes can also occur midplate, sometimes at ancient plate boundaries or at hotspots associated with volcanoes and magma movement. It is now also recognized that human-related activities such as mining, reservoir building, explosions, drilling, and wastewater injection can induce or create earthquakes.

Typically, when talking earthquakes, magnitude is used to describe the size, while intensity reflects the damage done. The intensity of an earthquake can vary based on location, the design and construction of buildings and other infrastructure, and the local geology. Magnitude, however, is not meant to vary over space or time; it reflects the amount of strain energy released. The popular press often expresses magnitude as "on the Richter scale," a measure developed in the 1930s by Charles Richter based very specifically on the amplitude of the trace of an earthquake on a Wood-Anderson seismometer located 100 kilometers from the epicenter. No one uses a 1930s-type Wood-Anderson seismometer anymore, and today most scientists use or refer to "moment magnitude" when discussing the size of an earthquake. This is a measure of how far the Earth's crust moved during a quake and how much force was required to move it, based on seismic recordings from multiple locations. (So forget "on the Richter scale" and just say magnitude or moment magnitude.)

Even with today's high-speed computers, the moment magnitude of an earthquake can take time to determine, so frequently a magnitude will be given as preliminary and will change slightly as additional data is processed. Magnitudes are reported on a base-10 logarithmic scale, meaning that for every whole number increase, the amplitude goes up by a factor of ten: A 7.0 magnitude earthquake is ten times stronger than a 6.0 magnitude earthquake, and an 8.0 magnitude earthquake is ten times stronger than a 7.0 magnitude earthquake (and one hundred times stronger than a 6.0 magnitude earthquake). In terms of energy, one step up in magnitude correlates to thirty-two times more energy being released.

A few more basics before moving on: The *hypocenter*, or focus, of an earthquake is the location within the Earth where an earthquake actually occurs, and the *epicenter* is the location on the surface above that spot. The *main shock*, the largest earthquake in a sequence of tremors, may be preceded by smaller *foreshocks* or followed by *aftershocks* that decrease in frequency with time.

Scientists now recognize another type of earthquake sequence, one in which there are numerous relatively small earthquakes. A larger earthquake may or may not occur, and if it does, it could be at the beginning, middle, or end of the series. These are called earthquake swarms. In 2017 an earthquake swarm struck Yellowstone that consisted of thousands of small temblors. Earthquake swarms have happened across the globe. Swarms may lead to increased stress on faults and another larger earthquake, or they may peter out to nothing; so far, there's no accurate way of forecasting which will happen—a frustrating unknown or wish-we-knew. Fluid or magma movement and/or the fracturing of rock may cause earthquake swarms.

Seismic waves—the vibrations released during earthquakes—have been the workhorse of earthquake science for years. They are how we detect, analyze, and locate earthquakes, and they have enabled us to learn a great deal about the Earth's interior. Seismic waves come in several varieties: P-waves, S-waves, and surface waves. P-waves, or primary waves, are compressional and move through the ground by jiggling molecules back and forth, parallel to the direction of travel. Stretch a Slinky and tap one end. A compression wave travels across the metal coil, moving sections back and forth. An important characteristic of P-waves is that their speed increases as the density of the surrounding material increases, and vice versa. P-waves can pass through both solids and liquids.

Now imagine lifting up or pushing down the end of a stretched Slinky. A wave again travels down its length, but the motion is due to changes in shape. S-waves—shear or secondary waves—propagate by deforming a material or shifting the molecules perpendicular to the direction of travel (from side to side or up and down). S-waves can pass through solids but not liquids, and they are slower than P-waves. Even slower are surface waves, which travel exclusively at the surface and tend to feel like a rolling motion.

The triangulation and analysis of seismic waves arriving at different seismic stations enables scientists to pinpoint an earthquake's location within the Earth and learn about what happened underground. While immensely helpful, this also highlights one of the greatest challenges faced in earthquake science: we can't directly observe what happens when a fault ruptures at depth.

Much like wanting to know what goes on in real time beneath a glacier or volcano, scientists dream of directly observing what happens underground during an earthquake. Since this remains outside the realm of possibility, at least for now, they must instead use indirect methods to unlock the Earth's secrets. However, as University of California seismologist David Jackson explains, indirect measurements require strong assumptions, which are, by necessity, made from our view at the surface. He wonders how many of our assumptions about earthquakes, stress, and the behavior of the Earth's crust at depth will turn out to be wrong, or at least not quite right.

Again, as in the study of glaciers and volcanoes, the history of instrument-recorded earthquakes is distressingly short—about 120 years. And as with volcanic eruptions, the largest earthquakes are the least frequent, so we know the least about the biggest events (though smaller eruptions and earthquakes can also cause disaster depending on where they occur). All of this makes it extremely difficult, and in some cases impossible, to understand the long-term history of earthquakes in a region or on a specific fault. Paleoseismology—the study of ancient earthquakes based on geologic deposits—is helping to extend our earthquake record. But we still have a long way to go.

Though many unknowns remain in earthquake science, huge advances in technology are providing scientists with increasingly precise and varied investigative tools. Seismometers, more sensitive and easier to deploy than ever before, can now be married with highly exact tiltmeters, pressure sensors, seismic and seafloor surveys, infrasound monitors (detecting frequencies below the range of human hearing), stunning visualization tools, and more. Seismic tomography, also used to study volcanoes, combines a number of techniques and allows scientists to create an image of the Earth's interior, essentially the geologic equivalent of a CAT scan. Researchers can also measure minute changes in elevation or horizontal movement using instruments linked to GPS or sensors aboard earth-orbiting satellites, like Interferometric Synthetic Aperture Radar (InSAR). Sophisticated laboratory experiments are used to assess how rocks and fluids behave deep in the Earth, and numerical and theoretical models allow us to simulate real-world events to investigate the processes involved and forecast what may happen in the future.

With recognition of the extreme dangers posed by earthquakes and tsunamis, and the certainty that they will happen again, nations such as the United States, Japan, Chile, China, Australia, and New Zealand have embarked on large-scale endeavors to use advanced technology to better understand the Earth and its underlying processes. The United States is at the forefront of this effort with its long-term, multidisciplinary program Earth-Scope, funded by the National Science Foundation. The effort includes three observatories (the USArray, the Plate Boundary Observatory, and the San Andreas Fault Observatory at Depth), hundreds of researchers, and the deployment of thousands of instruments. For instance, beginning in 2004, the USArray began a phased deployment of a permanent and transportable network of equipment across the United States and southern Canada. The deployment of seismometers was unprecedented in scope and density, and access to the data has been free for all interested parties (check out the cool animations of seismic wave motion at http://ds.iris.edu/ds/products/gmv/). For site-specific studies, a flexible array of more than two thousand seismic systems, along with a set of instruments that measure electromagnetic fields, have been made available. After rolling across the United States, the Transportable Array made its way to Alaska, the most seismically active state in the country. The EarthScope program and its partners have produced tremendous insights into the science of earthquakes, volcanoes, and the underlying Earth. Many of the projects described here have benefited from or have been part of the program (for example, the Imaging Magma Under Mount St. Helens project).

Ultimately, scientists want to better understand earthquakes in order to prepare and protect those at risk and provide more timely warnings. And despite the obstacles—earthquakes typically happen with little to no warning, and most of the action is underground—we've made great progress. Along the way, there have been surprises and some convention-cracking discoveries. Here are a few more game-changing events, what we've learned from them, and some wish-we-knews about earthquakes and tsunamis that continue to prove challenging.

Earthquakes: The Unknowns

The Landers Sequence

> Before this, no one thought faults could communicate. —**Ross Stein**, seismologist

California is a land built, modified, and frequented by earthquakes. In 1992 a series of events there unveiled new information about how earthquakes happen and progress, and raised many new wish-we-knews. They were not the most destructive, deadly, or headline-making earthquakes, but their impact on science was groundbreaking.

On April 23, 1992, a 4.0 magnitude earthquake and several smaller tremors struck the desert region of Southern California's Joshua Tree National Park. The earthquakes were worryingly close to the infamous San Andreas Fault. But when another earthquake hit later that day, it wasn't on the San Andreas as feared: it was a 6.1 magnitude event on a smaller fault to the north. It became known as the Joshua Tree event and was followed by a series of strong aftershocks.

Two months later and some 30 kilometers to the north, a cluster of relatively small earthquakes struck near the small desert town of Landers. The following morning Landers was hit again, this time by a major 7.3 magnitude quake. It was the strongest earthquake to strike Southern California in forty years, but because it occurred in a sparsely populated area there was little damage. The earthquake was shallow, only about a kilometer down. The fault rupture associated with it was some 85 kilometers long. But it wasn't just one fault that tore; the rupture unexpectedly triggered and ripped apart five separate faults, jumping between segments. This significantly increased the size of the quake.

Three hours after the Landers event and 40 kilometers to the west, a 6.5 magnitude earthquake occurred on the Big Bear Fault. Initially, it was described as a separate event or an aftershock, but scientists later realized the Landers event had triggered the Big Bear quake. Not only did the Landers earthquake unexpectedly jump between segments connecting separate faults; it also transferred enough stress to trigger another fault rupture some 40 kilometers away! Tens of thousands of aftershocks followed the Landers earthquake, and seismicity increased in areas as far away as Yellowstone. This was too far for the earthquakes to be considered aftershocks of the Landers event; these distant tremors were also something new—remotely triggered earthquakes.

After the Landers earthquake, seismic activity in the region eventually quieted down. But then, in October 1999, the Mojave Desert was hit again—by the 7.1 magnitude Hector Mine earthquake. Before the main shock, a cluster of smaller earthquakes occurred in the same area as a cluster of aftershocks from the Landers event. The fault rupture in the Hector Mine event was also essentially parallel to the Landers tear. Could the Hector Mine earthquake be strictly coincidence? Scientists no longer think so.

The series of earthquakes associated with the Landers event is now called the Landers Sequence—a series of cascading ruptures with unexpected jumps between faults that increased the magnitude of the event and triggered other earthquakes, some far away. It was a revelation for earthquake scientists and changed their way of thinking about how earthquakes happen and progress. The events also spawned numerous questions about how one earthquake triggers another. This continues to be a topic of great interest in the science of earthquakes. One possibility is that stress caused by an earthquake results in microscopic cracks that lead to failure on another fault. Another is that earthquakes cause fluids to migrate along a fault or within the crust, which then triggers another earthquake.

After the Landers Sequence, another worry arose. Could it have increased the stress on the nearby San Andreas Fault? Maybe, but how much stress or strain was already present—what was the starting point? This is another big unknown in earthquake science and a huge obstacle to forecasting earthquake occurrence and size. To make it even more complicated, recent research has revealed that earthquakes may or may not release all of the strain built up on a fault. There can be leftover strain that may contribute to a future earthquake. And even if the amount of strain present can be determined, scientists rarely know how much is needed to trigger an earthquake, or what size it will be. As our knowledge about earthquakes grows, so does the number of questions and wish-we-knews.

Cascadia: Disturbingly Quiet

In the Pacific Northwest, running alongside and off the coasts of Washington, Oregon, Northern California, and Canada's Vancouver Island is the 1,100-kilometer-long Cascadia Subduction Zone. Here, the Juan de Fuca Plate and smaller Gorda and Explorer Plates are moving east and sinking or diving beneath the North American Plate. The general configuration of the subduction zone is similar to those off Japan and Sumatra, but with some startling differences. Overlying the Cascadia Subduction Zone, one would expect a deep-sea trench. But charts of the seafloor show no such feature. In addition, for hundreds of years there have been no major earthquakes in the area; the subduction zone has lain eerily quiet. For years, the thinking was that either the converging tectonic plates are no longer converging—it's no longer an active subduction zone—or that the plates are creeping smoothly, either separately or in sync. Unfortunately, neither of those theories turns out to be correct: the Cascadia Subduction Zone is active and has produced giant megathrust earthquakes and destructive tsunamis.

To start with, there actually is a deep-sea trench overlying the Cascadia Subduction Zone, but it is filled with sediment. Using seismic surveys that can penetrate the seafloor to reveal the underlying structure, scientists have discovered that the narrow gorge typifying a deep-sea trench is, in Cascadia, chock-full of dirt. High rainfall events in the region, combined with several large river outlets, has resulted in the transport of huge quantities of sediment from the land to the ocean. Over time, this has filled in and hidden the trench overlying the offshore subduction zone.

As for the subduction zone being inactive, or creeping smoothly, that idea was first debunked in the 1980s, when geologist Jim Savage and his colleagues published survey data showing that in coastal Washington the ground was deforming—warping or bulging ever so slightly. The land was literally being squeezed together by the movement and sticking of the tectonic plates in the underlying subduction zone.

The presumption of a quiet and nonthreatening Cascadia Subduction Zone took another hit several years later when scientists made a surprising discovery. It played out like a geologic detective novel complete with a cliffhanger ending. The story begins in 1986 with US Geological Survey geologist Brian Atwater mucking around in the coastal marshes of westernmost Washington. Atwater, often armed with an old shovel, discovered buried soils and a series of sand layers going back a few thousand years. He also found a submerged expanse of cedar trees, still upright but long dead—dubbed a *ghost forest*. Atwater deduced that the old cedar trees and buried soils had submerged abruptly. Which raised the question: What could have caused such an abrupt submergence? The geologist theorized that if the locked tectonic plates within the Cascadia Subduction Zone released, it could cause a major earthquake and result in rapid subsidence at the coast that could immerse the trees and soil. And if the earthquake triggered a tsunami, it would explain the sand layers as well. But there was no historical record of such an event.

Scientist David Yamaguchi and his colleagues then used tree rings in combination with radiocarbon dating to show that something unusual had happened in the area about three hundred years ago. The story took an exciting turn when Kenji Satake and his colleagues found historical records of an "orphan" tsunami. The tsunami occurred in Japan in 1700, but they could find no evidence of its "parent," a corresponding and triggering large earthquake in Japan, South America, Alaska, or Russia. They eventually concluded that a giant earthquake in the Cascadia Subduction Zone was the best fit, and based on the data, they estimated that it had happened on January 26, 1700. Evidence suggested that the entire length of the subduction zone could have ruptured, creating a 9.0 magnitude monster. Satake and his colleagues also found Native American legends that told of a large earthquake occurring on a wintry night around that time. It all fit with Atwater's findings in western Washington. The moral of the story: the Cascadia Subduction Zone wasn't so quiet after all. It had produced giant megaquakes and colossal tsunamis, and could do so again. It was a startling wake-up call and an ominous warning for those living along the Pacific Northwest coast.

An exciting sequel to the story came out a decade later, when seismologist John Adams and his colleagues found a way to identify past earthquakes in the Cascadia Subduction Zone based on offshore geologic deposits. Within a series of sediment cores, they identified specific layers or deposits known as turbidites, the result of giant undersea debris flows presumed to have been triggered by large earthquakes. The scientists discovered a sequence of such deposits along a 600-kilometer stretch off the coasts of Washington and Oregon, which indicated that within the last 7,700 years at least thirteen big earthquakes had occurred in the Cascadia Subduction Zone.

Marine geologist Chris Goldfinger and his colleagues later extended this sequence of debris flow deposits to ten thousand years, identifying nineteen major earthquake events. The time between events appears to vary from two hundred to twelve hundred years. But the data also revealed something surprising: all of the earthquakes that were 8.7 magnitude or larger appear to cluster. Groups of four to five major earthquakes occurred some two hundred to five hundred years apart, with seven hundred to twelve hundred years between clusters. The largest earthquakes happened at the end of each cluster. These results fostered a whole host of new questions, including what causes the clustering and what does it mean for the region?

The phenomenon of earthquake clustering has been found elsewhere as well, including in the subduction zone off Japan. The reason for clustering, however, remains a giant unknown. Various theories have been proposed, focusing on some intrinsic characteristic of the fault, long-term processes or cycles related to strain or fluid movement within the subduction system, or synchronization by triggering (if a large earthquake triggers other large earthquakes it may essentially reset and synchronize the clock for strain accumulation and release). Then again, earthquake clusters could be a completely random phenomenon. Or the cause could be some as yet unrecognized external forcing factor—an unknown unknown. Recently, several scientists have proposed that clustering of large earthquakes on a global scale could be due to periodic slowing of the Earth's rotation. A test of this theory may be in the works, as a period of rotational slowing is expected in the next several years.

Goldfinger and his team made another startling discovery when they compared geologic deposits from the southern section of the Cascadia Subduction Zone to those found off the northern section of the San Andreas Fault. Thirteen of the fifteen debris flow deposits studied correlate in terms of time, suggesting that earthquakes in the southern portion of Cascadia occurred at almost the same time as those on the northern San Andreas, including the 1906 San Francisco quake. Goldfinger notes that it's either an amazing coincidence or one fault triggered the other. But he also points out that the information provided is inexact. The correlated events could be separated by hours or days, or it could be a matter of decades. Because Cascadia events tend to be larger than those on the San Andreas (a subduction zone versus a transform fault), the scientists concluded that large earthquakes in the southern reaches of the Cascadia Subduction Zone could trigger ruptures on the northern San Andreas.

They have also found, based on the data, that the energy released by earthquakes in the Cascadia Subduction Zone is not closely tied to a specific recurrence interval, or time between events. This means we don't know how much strain is present at any given time, nor do we know how much strain equals the breaking point. In other words, we still cannot forecast when the next big Cascadia event will occur, or whether it will hit in the south, north, or along the entire fault. But we now know it will happen, and at some point, it will be big, potentially catastrophic.

While the earthquake and tsunami hazards associated with the Cascadia Subduction Zone are now recognized and work is being done to prepare those at risk (such as improved tsunami education, identification of impact zones, evacuation route signage, and some reinforcement of school buildings), there's much more to be done. But it's an expensive proposition to reengineer structures and move existing or planned development. It is also hard to swallow such costs when a megaquake and tsunami may not happen for years to come. Finding the political will and investments needed to reduce risk is an ongoing and daunting challenge, but based on what we now know, it is more warranted than ever.

Random or Related

> We were surprised by the Tohoku event in Japan, and we were surprised by this earthquake. I think a little humility is good here. There are still a lot of unknowns in the planet, and we need to work a lot more. −University of Oregon seismologist **Diego Melgar,** who was raised in Mexico City

In 2017, within a mere sixty days, three large earthquakes struck Mexico. Was it pure coincidence, or were they related, perhaps a chain reaction? The first and largest earthquake hit on September 8 off the coast of Chiapas, in southern Mexico. It was an 8.1 magnitude temblor located some 47 kilometers below the surface, between the coast and the offshore Middle America deep-sea trench. Then on September 19, a 7.1 magnitude earthquake rocked the city of Puebla, at a depth of some 48 kilometers. The Puebla earthquake was ten times smaller than the previous quake, but given the local geology and closer proximity, the shaking wreaked havoc in Mexico City, 120 kilometers to the northwest. Hundreds of people were killed, more than forty buildings collapsed, and thousands of homes were damaged.

Mexico City is built on an old lakebed whose water-soaked clays can jiggle like Jell-O or liquefy in an earthquake. Tragically, it appears that the frequency of the jiggling on September 19 resonated with the natural sway of buildings eight to thirteen stories tall, causing unexpected collapses. To make matters worse, another earthquake struck just four days later. This time it was a 6.1 magnitude temblor striking 434 kilometers southeast of Mexico City, almost halfway between the two previous events. Together these three events highlighted the complexity of the subduction zone off Mexico's southern coast, the seismic hazards of the region, and some of what we still don't know but wish we knew about earthquakes.

Mexico is a seismic nightmare, one of the most active earthquake regions on the planet. For years, especially since the catastrophic 1985 earthquake that killed at least ten thousand people in Mexico City, scientists have been studying the nearby subduction zone. Just offshore, the northeastward-moving Cocos Plate is colliding with the westward-moving North American Plate. As a result, the Cocos Plate is diving, or being subducted, under Mexico. Like a deep gash sustained in the underlying tectonic battle, the Middle America Trench lies atop the offshore subduction zone.

But here, the plate boundary is exceptionally complex. The area hosts a triple junction, where the Cocos, North American, and Caribbean Plates intersect. Each plate is moving at a slightly different speed and in a different direction. The subduction zone is highly segmented, curved, and crossed by transform faults. Furthermore, the descending Cocos Plate begins its dive at a shallow angle, but then, some 300 kilometers inland and about 40 kilometers under Mexico, descends more steeply. The Cocos Plate itself also hosts a highly fractured zone, which is being drawn down and under southern Mexico and coincides with one of two seismically quiet zones in the area—the Tehuantepec and Guerrero Gaps. Neither of these segments has been hit by a major earthquake in more than a hundred years, or at least as long as records have been available. There's also a submarine mountain chain that is being pulled under Mexico. And Mexico is slowly crumpling upward, creating mountains to the south. It's like a mind-boggling 3-D jigsaw puzzle whose pieces vary in size, shape, and texture, move at different rates, can overlap and get stuck but also slip, and are mostly hidden from view.

Before the triple-earthquake whammy of September 2017, many people were looking to the two quiet zones as the most likely places to rupture next and cause a major quake. Particular attention has been paid to the Guerrero Gap because it is closer to Mexico City—now home to some twenty-five million people.

But surprisingly, none of the three 2017 earthquakes occurred in either the Guerrero or the Tehuantepec Gap, nor did they happen, as expected, on a plate boundary. The first two quakes were within the Cocos Plate below the plate boundary. Scientists think the bending of the descending slab may have caused this unusual type of subduction zone earthquake. The third, 6.1 magnitude earthquake was much shallower than the other two, and within the crust atop the overlying North American Plate. And that wasn't the only difference. The first, 8.1 magnitude quake was followed by numerous aftershocks directed upward and to the northwest. But the second, 7.1 magnitude quake had few aftershocks. Why the difference? That remains a seismic mystery. The bigger question is: Was there a connection between Mexico's three major shakers in 2017, and if so, what was it?

Scientists now believe the first two earthquakes were pure coincidence. They happened by random, though unlucky chance. The second earthquake that did so much damage in Mexico City was also eerily coincident with the deadly 8.1 magnitude 1985 earthquake, happening on the very same date thirty-two years later. It also occurred just two hours after Mexico City's annual earthquake drill.

What about earthquake number three, the 6.1 magnitude event on September 23? This earthquake occurred just northwest of the aftershock sequence from the first and largest earthquake, the September 8 event. Scientist Shinji Toda of Tohoku University, Japan, calculated the stress created by the earlier earthquake and its aftershocks. He concluded that the first earthquake increased strain on the fault that ruptured in the third quake—so there was a connection.

The events of September 2017 highlight some lingering questions about Mexico's seismicity and the next big earthquake. Should scientists focus their monitoring efforts on seismic gaps along the subduction zone, where strain has built up over long periods of time without release, or places where earthquakes have happened before and may be more likely to happen again? Episodes of slow-slip and related tremors (more in the next section) have now been identified in association with the Guerrero

Gap. What role do they and the subducted fractures and ridge play in Mexico's complicated subduction zone and pattern of earthquakes?

Scientists from Mexico and Japan have teamed up to try to answer some of the questions posed by Mexico's tectonic puzzle. An extensive network of instruments that include GPS, pressure, and seismic sensors is being deployed off Mexico's southern coast. Over the next four years researchers will be collecting data and closely monitoring the area, hoping to learn enough to help mitigate the risks from earthquakes in Mexico and elsewhere.

It may turn out that each of the world's subduction zones, much like volcanoes and glaciers, behaves differently. One recently discovered phenomenon, known as a slow-slip or silent earthquake, appears to occur in some subduction zones but not all, and can vary when it does occur. Tantalizingly, slow-slip events have preceded great earthquakes in Japan, Turkey, Chile, and Mexico. Could they help us actually achieve the dream—the dream of forecasting megaquakes?

Slow-Slip or Silent Earthquakes

> Herb was very worried at first—he thought something was wrong with the data. He tried everything to prove himself wrong, and everything failed. –**Kelin Wang**, who worked with Herb Dragert and Thomas James to solve the slow-slip mystery

Some twenty years ago an astounding discovery radically changed our understanding of how strain accumulates in the earth and is released. In 1999, while reviewing data from GPS monitoring stations in southwest British Columbia, Canada, and northwest Washington State, geophysicist Herb Dragert noticed something strange. The motion in a cluster of seven sites had briefly and unexpectedly reversed direction. With the locked Cascadia Subduction Zone just offshore, these stations typically showed landward movement as the converging tectonic plates squeezed and deformed the ground. But the GPS data revealed the stations had moved seaward, as if the tectonic plates had momentarily un-locked and rebounded—as happens in an earthquake. But there had been no shaking, nor were any seismic waves detected indicative of an earthquake.

Dragert and his colleagues examined the data more closely to see if it was faulty; maybe something had gone wrong with the equipment. It wasn't the instruments. They eventually confirmed that an area 50 by 300 kilometers had slipped seaward on average about 2 centimeters. It was an offset that might be expected during a 6.7 magnitude earthquake. But instead of happening in an instant, this offset occurred over a period of some six to fifteen days—it was a slow slip. So slow it was essentially silent; it had produced no discernable seismic waves. The team further determined that this gentle sliding had occurred some 25 to 40 kilometers below the surface, in an area below the locked portion of the Cascadia Subduction Zone where earthquakes usually happen.

At about the same time, seismologist Kazushige Obara was examining data from Japan's high-sensitivity seismograph network (Hi-net), which is designed to detect microquakes. He discovered deep long-period, low-amplitude vibrations or tremors occurring in the subduction zone off southwest Japan. The tremors occurred at an average depth of about 30 kilometers—around the same depth as the slow slip in Cascadia. Some of the tremors lasted minutes, others several weeks. The vibrations resembled what is known as volcanic tremor, which is indicative of magma moving underground. But there was no volcano or magma in the region. Later it was found that slow slip is often, but not always, accompanied by such tremors. Scientists Garry Rogers and Herb Dragert named the phenomenon episodic tremor and slip.

With the availability and deployment of increasingly precise instrumentation, slow slip and tremor have now been documented across the world, in subduction zones off Alaska, Mexico, New Zealand, Costa Rica, and Japan. Slow slip can also happen without tremors and can vary in depth and periodicity. In the Cascadia Subduction Zone, slow slip and tremor have been found to occur on a remarkably regular basis, about every fourteen months, to last for up to five weeks, and to cause about 5 millimeters of displacement. Slow slip has also been identified on transform faults such as the San Andreas.

Unlike typical earthquakes, which release strain nearly instantaneously, slow slip releases stress on a fault over a period of days, weeks, or even months. While stress may be reduced on one portion of a fault, slow slip may add stress to other regions, including at the base of the locked portion of a subduction zone, where megaquakes are generated. This is why instances of slow slip before a major or great earthquake are so thought-provoking: do they increase the likelihood of, or could they in fact trigger, a big quake? Could slow slip one day be used to forecast or warn of an impending megaquake? The answer remains unknown.

In the subduction zone off Costa Rica, scientists have discovered that slow slip can happen at various depths and that, while it may not be predictive of megaquakes, it could help to assess the magnitude of future earthquakes and their potential to generate tsunamis. To better understand and monitor slow slip off Costa Rica and in other subduction zones, researchers are working to develop new technologies to measure strain offshore. The University of South Florida's Tim Dixon is developing buoy-based technology to monitor strain in the shallow offshore region, while other groups are working on deep-water systems. He notes that we're probably five to ten years away from having technology adequate to cover all the real estate, from trench to coast.

Now that we know slow slip happens, a big unknown is why? What is the cause, deep in a subduction zone or on any fault, for that matter? In subduction zones one plausible explanation is that the release of water at depth could reduce friction and allow the plates to slip more smoothly.

Slow slip and tremor are now part of earthquake lexicon and have changed our understanding of earthquake processes and subduction zones. Such discoveries, especially ones that arise through advances in technology, make one wonder what will be uncovered next?

Puzzling Subduction Zones

Given their propensity to host the planet's strongest earthquakes and trigger devastating tsunamis, the world's subduction zones are under intense scientific scrutiny. Large-scale, cross-disciplinary research efforts are under way in the United States, Japan, Mexico, and Chile. Some of the challenges are by now familiar. With the ocean and earth obstructing our view, yet again we cannot directly observe what is going on in a subduction zone. It is expensive and logistically difficult to deploy instruments and maintain them at critical sites, in this case, thousands of meters underwater. And again, the available data represent only a short time; our instrument observation record goes back a little over a hundred years, while some earthquakes may have recurrence intervals of hundreds or thousands of years.

Scientists have long hoped to find some property of subduction zones that could be used to predict the maximum magnitude of earthquakes they might generate. Factors such as the width and age of the descending slab, the rate of descent, the amount of sediment in the overlying trench, the rate of strain accumulation, and the curvature of the trench or plate boundary have been considered. So far, the jury remains out, as none of these or other properties have been shown to correlate well with maximum earthquake size. One promising factor is the angle of descent in the downgoing tectonic plate or slab. If the slab descends steeply, less surface area is involved and it may heat up more rapidly, becoming less brittle and therefore less likely to sustain larger earthquakes. But even this relationship is unsure.

Unknown Faults and Connectivity

It's been described as one of the most complex earthquakes ever recorded. It tore through mapped and unmapped faults, caused extensive uplift, a tsunami, and widespread damage. This earthquake literally ripped apart conventional wisdom about rupture processes and segmented faults.

NEW ZEALAND, 2016. Just after midnight on November 14, a 7.8 magnitude earthquake struck about 90 kilometers northeast of Christchurch, New Zealand, in the coastal town of Kaikoura. Shaking was felt throughout the country, but most of the damage was on South Island, and a three-meter tsunami struck Kaikoura. To better understand what had happened, scientists took to the field to collect data and analyze information from a wide array of instruments. What they found once again surprised even the experts.

The fault rupture started in the south and propagated northward for more than 170 kilometers. It tore along known and previously unknown faults and then spread offshore. In total, some twenty faults ruptured. While it was known that ruptures could jump between faults (the Landers Sequence), it was thought that the distance for rapid triggering was limited to about 5 kilometers. Here, some of the faults or segments of faults were separated by as much as 20 kilometers, and the triggering between faults was nearly instantaneous. The event also caused slip in the nearby offshore subduction zone and up to 12 meters of ground deformation.

The Kaikoura earthquake made scientists rethink how models simulate fault rupture and how we forecast maximum earthquake potential and related hazards. It also brought attention to how the earthquake would have been interpreted if it had occurred in the distant past and been identified through paleoseismic techniques. Each fault rupture may have been documented as a separate earthquake instead of a part of the overall event. And it highlighted unanswered questions, for instance, what controls the rupture process, from beginning to end? And how many faults remain unknown, especially near or below highly populated areas?

"Christchurch has never been identified as a major earthquake zone, because no one knew this fault ran beneath," noted Roger Musson, a seismologist with British Geological Survey. And 2016 was not the first time, even in well-mapped New Zealand, that earthquakes had revealed the presence of unknown faults. In 2011 a shallow 6.3 magnitude earthquake struck around lunchtime just 10 kilometers from Christchurch. The violent shaking caused buildings to collapse, thousands were injured, and at least 185 people died. Thousands of earthquakes rock New Zealand each year, as it lies on the boundary between the Pacific and Australian Plates and adjacent to two subduction zones. Yet the fault on which the 2011 earthquake occurred had only been discovered the previous year during another earthquake. Seismologists estimate it may not have ruptured in thousands or tens of thousands of years, so the city of Christchurch was built with no knowledge of the underlying fault and the risks it posed. Around the world there are plenty of other locations facing similar and potentially tragic dilemmas.

The 7.0 magnitude earthquake that struck Haiti in 2010, killing more than two hundred thousand people and leaving more than a million homeless, happened not, as was originally thought, on the well-recognized Enriquillo Fault, but on a roughly parallel, previously unknown fault—the newly discovered and named Leogane Fault. Research has also revealed that the Enriquillo Fault did not release its accumulated strain during the event; this means it is still primed for a major break.

Most of the world's known faults are on the Earth's surface. Many others, however, are hidden at depth and have yet to be recognized. Even in the United States new faults are still being discovered. Following a 4.5 magnitude earthquake in 2011, new faults were discovered near St. Louis, Missouri. The same year, previously unrecognized faults were documented in Virginia as a result of aftershocks from the 5.8 magnitude temblor that damaged the Washington Monument. And in California, new faults were discovered after the 2014 Napa earthquake and again in 2016, when scientists uncovered an offshore fault to the south of and roughly parallel to the San Andreas. The United States has some of the best fault-mapping technology in the world. Imagine how many more unrecognized faults exist in other parts of the world!

Human-Induced Earthquakes

It is now accepted that human activities such as reservoir im-poundment, mining, withdrawal of fluids and gas, and injec-tion of fluids into the ground can trigger earthquakes. These are considered induced events and occur mostly in relation to deep wastewater disposal. In 2011 a 5.6 magnitude earthquake oc-curred in central Oklahoma, destroying fourteen homes and in-juring two people. The Oklahoma Geological Survey released a statement reporting that the seismicity rate in the state at that time was six hundred times greater than background levels, and that the increase was very *unlikely* the result of natural processes. The US Geological Survey now includes induced earthquakes in its seismic hazard mapping. States at greatest risk include Arkansas, Colorado, New Mexico, Ohio, Oklahoma, Texas, and Virginia. Most induced earthquakes are fairly small, less than a magnitude 4.0. But there's a sizable unknown here: How big can induced earthquakes be? The answer remains to be seen or, more precisely, felt.

What Most People Wish They Knew

There remain many unknowns and wish-we-knews regarding earthquakes, such as how faults rupture at depth, why quakes sometimes cluster, and what causes a small earthquake to go big. But for most people, the number one concern is: When and where will the next temblor strike? And how big will it be? Some sci-entists believe earthquakes are so chaotic and the processes in-volved so complex that we will never be able to accurately predict them. Many theories have been tested, involving a wide range of possible precursors—from animal behavior to geomagnetic anomalies, groundwater changes to full moons and high tides—but none has proven reliable. Seismologist Susan Hough notes that in the 1970s seismologist Ruth Simon even devised experi-ments to see if cockroach activity could be correlated to impend-ing earthquakes. It couldn't.

We know earthquakes are going to rattle the planet, we have a pretty good idea where they are most likely to happen, and we know that some events are going to be megaquake monsters. With this knowledge there are ways to reduce our risks. Marine geologist Chris Goldfinger puts it this way:

> To make earthquakes and tsunamis manageable, we need education above all else. From education comes preparedness. From a place of preparedness, we find that prediction is really not that important after all. If we could predict, but were unprepared, how much good would it do? In a few cases, warning helps. . . . But the value of preparedness is that it doesn't depend on any new and untested model, nor on any device that may fail in the heat of the moment.

In his book *Earthquake Time Bombs*, seismologist Robert Yeats identifies heavily populated areas of the world that are at great risk for large and destructive earthquakes. It turns out that many of the world's megacities lay close to or on active faults and although large earthquakes may have happened there in the past, current residents may not have experienced an earthquake in their lifetime or even their parents' lifetime. Moreover, the infrastructure in such megacities may have been constructed before relevant information about seismic hazards came to light. Yeats's list of "earthquake time bombs" waiting to explode is scarily long and includes Kabul, Afghanistan; Tehran, Iran; Caracas, Venezuela, and Istanbul, Turkey. Other cities and areas at high risk can be found in Pakistan, India, Kenya, and China—and of course, all along the Pacific Northwest and in California.

Internationally, efforts are under way to provide more countries and regions with the means to assess earthquake risk, to improve readiness, and to examine the economic impacts of structural retrofits and relocation. These include the public-private partnership GEM (Global Earthquake Model) and the GEAR model (Global Earthquake Activity Rate). In some places, like

Istanbul, California, and the Pacific Northwest, steps are being taken to reduce risk. In other areas, especially in developing countries, little is being done to prepare for the inevitable.

The dream of predicting earthquakes remains, but for now scientists are working hard to improve our ability to forecast the likelihood of occurrence, determine the ground motion that may occur, and identify potential impacts. Based on the best available science and sophisticated computer rupture models it is estimated, for instance, that in California there is a 93 percent probability (a very, very good chance) of an earthquake of 7.0 magnitude or larger occurring by 2045, with the highest probability for it to occur on the San Andreas Fault. National seismic hazard and potential ground-shaking maps are available through the US Geological Survey. Unfortunately, many people find earthquake probability forecasts either confusing or unconvincing. In response, some scientists believe we should drop probability forecasting and focus on the fact that large quakes are going to happen and that some areas are at greater risk than others—earthquake time bombs.

California's population now stands at some 40 million people. In 2008 the US Geological Survey released "The ShakeOut Scenario," a report on what would happen should a 7.8 magnitude earthquake strike along the southernmost end of the San Andreas Fault. The estimated impacts were eye-opening and distressing: eighteen hundred fatalities, fifty thousand injuries, and a cost of $200 billion. In response, California has since held an annual earthquake drill, the Great California Shakeout, to better prepare communities and residents. Along with improved preparedness, some countries, including Japan, Taiwan, Mexico, and Romania, have operational early-warning systems. Similar systems are under development or in the early stages of implementation in California and the Pacific Northwest. But drills and an early-warning system are not enough; it is equally important in high-risk areas that new buildings are earthquake-resistant and old structures are retrofitted.

Seismologist Ross Stein likes to conduct a simple yet startling demonstration. He shows two stacked-cube structures made of sticks, one with bracing at the corners and one without. When he shakes them, simulating an earthquake, the structure without corner braces fails quickly while the other only sways slightly. We know how to make buildings safer in earthquakes; the solution is not complex. The problem isn't the earthquakes or not knowing what needs to be done. The problem is having the political will and funding to do it.

"During an earthquake is not the time to wonder if your building is safe," says California assemblyman Adrin Nazarian. In 2018 he introduced a bill in the California legislature to require each city or county to create a public inventory of seismically vulnerable buildings. The idea was to identify buildings at risk of collapse in an earthquake and then use market forces rather than statutory mandates to drive retrofits. Many people don't realize the buildings they live or work in would be unsafe in an earthquake. Or they may think everyone in an earthquake zone is at the same level of risk. They're not. If you live or work in an older or unreinforced building your risk may be much higher. On the economic side, estimates are that for every dollar spent on retrofits, some four to seven dollars are saved in repair costs. Nazarian's bill did not, however, pass.

Even in the often-contentious science community, and with plenty of earthquake wish-we-knews remaining, everyone agrees on one thing: we must do more in terms of earthquake readiness, planning, education, and mitigation, especially in areas of high risk.

Tsunamis

In Sumatra in 2004 and Japan in 2011, the world witnessed the devastation caused by tsunamis. From these events, we learned that it is not just the size of a triggering earthquake that matters; complexities in the rupture process can cause a moderately sized earthquake to generate a larger-than-expected tsunami. We also learned that bathymetry can steer a tsunami's energy along a specific path. It's not exactly like dropping a pebble in a pond—sometimes energy gets directed by the underlying mountains and valleys of the seafloor. An immense amount of knowledge has been gained from those two events, but it came at a tragically high price.

Scientists continue to research and model tsunamis, and every time a tsunami strikes our understanding grows. But the study of tsunamis is again difficult. They happen even less frequently than earthquakes and cannot be predicted in advance. One thing the experts agree on is we simply don't have enough data. Researchers are trying to add to the pool of data on tsunamis by searching for geologic evidence of prehistoric events, but it is a tricky endeavor. Much of the evidence has been eroded, washed away in subsequent events, or buried beneath newer deposits.

The range of technology available to researchers is rapidly advancing, from satellite-based sensors to undersea pressure sensors and large wave tanks in which scaled-down versions of a tsunami can be created and observed. Still much of our understanding comes from observing real-world events.

The Pacific Ring of Fire, lined by subduction zones, is the birthplace of most of the world's tsunamis. Sitting smack-dab at its center, like a bull's-eye in the tsunami target zone, is Hawaii. And of all the places in the Hawaiian Islands that have been hit, one town is the unlucky standout. On the northeast shore of the big island of Hawaii sits Hilo, fronted by a shallow semicircular bay and breakwater. The building of commercial and residential structures directly on Hilo's waterfront is now prohibited, and with good reason. In 1946 and again in 1960 the town and its waterfront were hit hard by tsunamis.

Early in the morning on April 1, 1946, a monster subduction zone earthquake, now estimated at 8.6 in magnitude, struck 145 kilometers from Unimak Island in the Aleutian Islands. The megaquake triggered a tsunami that sped out across the Pacific Ocean. Within minutes, a wall of water some 42 meters high struck nearby Alaska. The lighthouse on Unimak Island was destroyed and five people were killed.

Meanwhile, the tsunami raced south across the Pacific. Flying over the ocean, a Navy pilot on patrol noticed an odd line in deep water. After calling his base in Oahu, he was ordered to reduce altitude and investigate. The pilot reported startling news: the line was moving faster than his airplane. It was the wave train of the tsunami speeding across the open sea.

Some five hours after the earthquake and without warning, the tsunami arrived in Hawaii. Waves reportedly reached up to 16 meters high in some areas. Due to the shape of Hilo Bay, the tsunami was amplified, and it caused severe damage to the densely developed waterfront. The majority of the buildings were swept away or destroyed, and the rail system and port heavily damaged. Survivor Dan Nathanial described the tragic event, starting with the characteristic retreat of water from the bay:

> The ocean bed was practically dry. You could see the bottom. . . . [Then] the first wave came in . . . filling the ocean bed. You could see the current race in the direction of Waiakea, hit Sea View Inn, and then swing toward Coconut Island. . . . It smashed the bridge and went on its way of destruction. When the next big wave came, I was standing in front of Hilo Meat Co. The next thing I knew I was hanging to the rafters for dear life, with boiling water all about me.

A total of 159 people were lost in Hawaii, 96 in Hilo alone. Adjusted to 2011 dollars, the estimated cost of the disaster was $300 million. As a result of the disaster, the United States established the Seismic Sea Wave Warning System.

Hilo again bore the brunt of a massive tsunami when the largest earthquake ever recorded struck off the southern coast of Chile. It was May 22, 1960, and the earthquake was a mega 9.5 magnitude subduction zone behemoth that literally rang the Earth like a bell and triggered a powerful tsunami. In Chile, thousands of people died and an estimated two million were left homeless. It took fifteen hours for the tsunami to reach Hawaii. Hundreds of people were injured and sixty-one killed. Hundreds of homes and buildings were also damaged or destroyed. And as in 1946, Hilo was especially hard hit. Multi-ton boulders used to build Hilo's breakwater were picked up and dumped on downtown streets. After ravaging Hawaii, the tsunami went on to strike Japan, killing another two hundred people.

Following the 1960 tsunami, nations around the Pacific came together with one goal in mind: to prevent loss of life due to tsunamis. The Seismic Sea Wave Warning System soon became the Pacific Tsunami Warning System. Dedicated facilities were later established in Hawaii, on the US West Coast, and in Alaska, which together were eventually renamed the National Tsunami Warning Center. The tsunami warning system is now an international effort with twenty-six members. When a potential triggering event occurs, such as a coastal earthquake of 7.0 magnitude or greater, tsunami warnings are provided to communities throughout the Pacific and Indian Oceans, South China Sea, US Atlantic and Gulf of Mexico coasts, Canadian Atlantic coast, and the Caribbean. But even with a warning system in place, some tsunamis still come as a tragic surprise.

NICARAGUA, 1992. A loud rumbling had accompanied previous earthquakes in the region, and when the ground shook coastal residents knew to go to high ground. But on September 1, 1992, an earthquake occurred off Nicaragua without any rumbling, and the shaking was so weak few people noticed. Some twenty minutes later, a giant wall of water up to 10 meters high struck Nicaragua's Pacific coast. The ocean raced ashore and went more than a kilometer inland, washing away homes and causing some two hundred fatalities. More than 13,500 people were left homeless.

The experts were stunned. The early estimate was that it had been a 6.8 magnitude shaker. That was below the 7.0 magnitude threshold that triggers an automatic tsunami warning. Later, with further analysis, the magnitude was revised to 7.7; it had been a much stronger earthquake than originally thought. What caused the underestimation? Why hadn't there been notable shaking? And why did the earthquake trigger such a large local tsunami?

The temblor that hit Nicaragua that day was what is now known as a "tsunami earthquake." Scientists had previously known about tsunami earthquakes, but until September 1992, they had never seen the seismic record of one or had enough information to develop a way to recognize it quickly. In a tsunami earthquake the rupture process is slow, the release of energy gradual, and the duration generally longer than a typical earthquake. There is little to no shaking, and the short-period seismic waves that arrive first at a seismometer do not necessarily provide an accurate indication of the earthquake's true magnitude. In the 1992 Nicaragua quake, the short-period P-waves and surface waves were initially used to calculate the magnitude (6.8). But later, when researchers factored in longer-period S-waves, the true magnitude of the tectonic beast was revealed (7.7). The event is considered a watershed moment because it was the first tsunami earthquake ever recorded by a modern broadband seismic network.

In addition to lasting almost two minutes, the 1992 Nicaraguan earthquake was shallow and ruptured all the way up to the seafloor. Research suggests that this and the presence of soft sediments in the interface between the Cocos and underlying Caribbean Plate contributed to the slow rupture process and generation of a larger than expected tsunami.

In part due to the unusual nature of the earthquake, an international team of scientists was sent to Nicaragua to conduct surveys in the stricken region. It was a first in international collaboration and led to the regular deployment of an International Tsunami Survey Team in response to such events. Since 1992 international teams have responded to more than thirty-three tsunamis.

The 1992 Nicaraguan event showcased the benefits of modern seismic detection and analysis capabilities, but also the difficulty of providing tsunami warnings when the triggering earthquake occurs nearby and is of the slow, long, and nonshaking variety. US Geological Survey geophysicist Eric Geist laments that all of our knowledge about earthquake mechanics doesn't seem to explain tsunami earthquakes, but they are extremely dangerous. Some scientists suggest that an undersea landslide may have accompanied the 1992 earthquake and helped generate the tsunami. And here is another big unknown in tsunami science—undersea landslides. We know they can trigger tsunamis, but we have no way of detecting them in real time or of predicting them. Another disaster some six years after the Nicaraguan event highlighted the problem.

PAPUA NEW GUINEA, 1998. It was July 17 and along the shores of Sissano Lagoon, villagers had gathered to prepare the evening's meal and celebrate the start of a four-day national holiday. Just before 7:30 p.m., a loud bang reportedly echoed across the region. The earth began to shake. But earthquakes were common here. To the locals, the shaking that night wasn't out of the ordinary or something to be concerned about. A few moments later the ocean began to recede, and soon after a wall of water, some 10 meters high, rushed ashore.

Three tsunami waves hit Papua New Guinea that night, each 7 to 15 meters high. Seawater rushed over the narrow stretch of sand fronting Sissano Lagoon and the nearby fishing villages. Most of the houses were swept away and some three thousand people killed or lost. It took hours for the outside world to even learn of the calamity.

Initially, scientists thought an offshore earthquake had generated the tsunami. But the quake's magnitude was estimated at 7.1, and experts questioned how it could have triggered such an enormous tsunami. Based on data collected after the event by the International Tsunami Survey Team, scientists came to suspect that the earthquake was not to blame for the tsunami. It had been an undersea landslide.

The event again highlighted how quickly tsunamis can strike locally and brought greater attention to the idea of earthquake-induced landslides contributing to tsunami generation. At the time, tsunamis triggered by undersea landslides were still considered rare. This despite the fact that in 1958 an 8.0 magnitude earthquake struck Alaska and caused a glacier to avalanche into a narrow fjord in Lituya Bay. The landslide triggered a tsunami that reached a whopping 457 meters high—the tallest ever recorded.

Scientists now think that landslides into or under the sea have contributed to many of the tsunamis that have occurred over the past several decades. This was again illustrated in 2018 by two destructive tsunamis in Indonesia, one in Palu, which evidence suggests was caused by a submarine landslide triggered by an earthquake, and the other in the Sunda Straits, generated by flank collapse during an eruption of the Anak Krakatau volcano. How often does this happen? We wish we knew. Scientists are working hard to better understand and find ways to detect and warn of landslide-generated tsunamis.

Tsunamis: A Few Basics

Much about tsunamis can be gleaned from the previously described events. In short, tsunamis are most often caused by large displacements of water due to the up or down motion of the sea-floor. Triggering events include earthquakes, landslides, volcanic eruptions, or asteroid impacts. Once water is displaced, the causative energy is transferred into motion in the form of waves. In the open ocean, these waves are long, low, and fast—moving at about 800 kilometers per hour. Unlike wind waves, which tend to travel only in the upper part of the ocean, tsunamis affect the entire water column, even in very great depths. At the surface in the deep sea, however, a tsunami can pass by unnoticed. Tsunamis can also travel great distances with little energy loss.

When a tsunami approaches the shore and depth shallows, the waves begin to "feel bottom" and slow at their base due to friction. The upper part, however, continues undisturbed, causing the wave to "bunch up," or steepen and grow in height. At the shore, tsunamis slow, traveling at about 30 to 40 kilometers per hour—still too fast to escape from on foot. The height and shape of a tsunami varies depending on numerous factors including, the magnitude of the triggering event, the bathymetry it passes over, and the shape and nature of the coastline. Tsunamis can come ashore as a series of large waves or as surges, more like a superfast incoming tide. Coral reefs can cause a tsunami to break before it strikes the coast, while wetlands and mangroves can dampen the energy once it comes ashore. As tsunamis hit the shore, they create powerful currents that push inland. The seawater pushed ashore then flows swiftly and dangerously back to sea.

Forecasts and Warnings

If you're on a coastline and experience a strong tremor that lasts over 30 seconds, or if you observe any unusual water motions, move to high ground. **–Costas Synolakis**, tsunami expert, University of Southern California

Scientists are working hard not only to understand tsunamis but also to model them accurately and provide as much advance warning as possible. Typically, when an event occurs that could generate a tsunami, a warning is automatically triggered. In addition to the size of an earthquake, scientists take into account the history of earthquakes and tsunamis in the area. For a warning to be effective, especially for communities near a triggering event, it must come quickly and reach those at risk. And then, people receiving the warning must know what to do and where to go to be safe.

Models of tsunami generation, travel, and impacts are rapidly improving. Typically, the predicted time of arrival is right on target. But it remains difficult to forecast what shape and size a tsunami will take when it strikes. Part of the challenge is that tsunamis can behave in unexpected ways. As very long waves they can reflect off seawalls and create resonating waves that slosh back and forth in bays and harbors (a seiche). Tsunamis can also refract around islands and amplify, striking areas thought to be sheltered from danger. And where a bay narrows, it can act like a funnel and cause a tsunami to grow to enormous proportions. It also remains unclear why some tsunamis become monsters and others peter out.

An important aspect of tsunami warnings is confirmation; being able to confirm that a dangerous wave has in fact been triggered. Observations and tide gauge data are used to verify if and when a tsunami has struck land. But offshore it's more difficult. Near areas of high risk, there's a sparse but growing network of specialized DART (Deep-ocean Assessment and Reporting of Tsunamis) buoys. Typically placed in the deep sea, DART buoys include sensors on the seafloor that measure changes in pressure as waves pass overhead. Data is then transmitted to the surface where it is passed on via satellite to scientists monitoring the system. The data is filtered and analyzed in order to distinguish a possible tsunami from other waves passing by. If a potential triggering event occurs, the buoys are put into a rapid fifteen-second recording mode and closely monitored. But not only do DART buoys need to be in the right place at the right time, they also need to be working. The ocean is a harsh environment, and maintaining equipment is logistically difficult and costly. Sadly, DART buoys and other monitoring devices in the ocean often end up in disrepair.

Accurate modeling and forecasting of where a tsunami will strike and what shape it will take requires information, and a lot of it. As the 2004 Sumatra tsunami revealed, one big chunk of information needed is ocean bathymetry—the highs and lows of the seafloor that can amplify, beam, or dampen tsunami energy. Unfortunately, only about 5 percent of the seafloor has been mapped in detail. We have more detailed maps of Mars and the moon than of most of the ocean floor. Most bathymetric maps, including those found in the popular Google Earth, are based on satellite data (altimetry). The resolution of such maps is on the order of 5 kilometers, so features that are smaller go undetected. Higher-resolution mapping is expensive and requires the use of ship-based surveys and remotely operated vehicles. More detailed surveys, however, are a regular full-on festival of discovery, as noted by scientists Larry Mayer and John Clark, who've found that the seafloor continually offers surprises: new mountains, new volcanoes, the world's largest river channels created by undersea debris flows. Still, most of the seafloor remains a mystery or very vast unknown.

Scientists are also investigating, using numerical models and large wave tanks, how tsunamis change in response to and impact different types of structures. Another big question here relates to climate change: How will rising sea levels change the nature and impacts of tsunamis? "We don't understand well enough how much a half-meter change in sea level will enhance tsunami inundation," explains Costas Synolakis. "Yet we know well that our current flooding predictions will be entirely obsolete in less than two decades."

Through scientific research and observations, we will continue to learn about earthquakes and tsunamis, and with improved understanding will come the ability to better prepare and warn those at risk. To learn more about earthquakes and tsunamis, their history, related hazards, risk mitigation, and research, see the list of readings and references at the back of this book.

4

Hurricanes

AFRICA, AUGUST 2017. Waves of heat shimmered atop the desert sands of North Africa. To the south, it was cooler, wetter, in some areas forested. The differences in temperature and moisture between the two regions gave rise to an area of low pressure and storminess. Soon it was picked up and transported by westerly-blowing winds. The area of disturbed weather, now called an easterly wave, traveled off the coast of Africa and over the eastern Atlantic Ocean. There, rain clouds started to form, but they remained disorganized.

Over the warm ocean, the easterly wave was like a seedling waiting for the right conditions to grow and blossom—into a true storm. For the moment, though, conditions weren't ripe for growth, and as the wave or low pressure continued west, it dissipated. Then the ocean grew warmer. Rising air brought moisture evaporated from the sea upward, where cool temperatures caused the moisture to condense and form clouds. But up above, the winds shifted and changed with altitude—wind shear. The clouds broke up and the low pressure remained weak as it headed further west toward the Lesser Antilles Islands and Caribbean.

Overnight, conditions changed. The wind shear relaxed. Air

rising over the warm ocean now drove moisture skyward forming clouds and fueling the growth of thunderheads. Energy and moisture were released, and the density of the rising air decreased, causing it to march ever higher. The thunderheads grew taller. As more air rose, more was sucked in from below. Surface winds began converging toward the low pressure. Air continued to rise, and at the surface atmospheric pressure dropped. Due to the Earth's rotation, the strengthening winds started to spin counterclockwise. It was August 17, and some 815 kilometers east of Barbados, the easterly wave born off the African coast had become a tropical depression.

The tropical depression continued to move west as the warm ocean fueled its growth. More moisture rose into the atmosphere, the burgeoning clouds grew taller, and the winds intensified. The central pressure of the storm dropped further and late in the day, when its sustained winds topped 34 knots (63 kilometers per hour), Tropical Storm Harvey was born.

Harvey passed over Barbados and St. Vincent, bringing heavy rains and strong gusty winds. Once in the Caribbean Sea, however, Harvey encountered upper-level dry air and northerly wind shear. The tempest weakened. On August 19, the National Hurricane Center ceased its advisories. But Harvey wasn't done.

In the following days, the remnants of Harvey moved rapidly west and northwest, and on August 22 crossed over Mexico's Yucatan Peninsula and into the Gulf of Mexico. The ocean here was exceptionally warm, averaging about 29°C, about two degrees above normal. The unusually warm water fueled growing thunderstorms, atmospheric pressure dropped, and by August 23 Harvey was once again a tropical storm. The environment surrounding the storm was now ripe for strengthening: light wind shear, very warm ocean water to great depths, and a moist atmosphere.

Intensifying, Tropical Storm Harvey turned northwest, steering around the western edge of a distant subtropical high-pressure ridge. On August 24, its winds exceeded 64 knots (119 kilometers per hour); the storm became Hurricane Harvey and, still strengthening, made a beeline for the Texas coast.

What happened next surprised even the experts. Within the

span of forty-eight hours, Harvey went from a tropical storm to a major Category 4 hurricane (with winds greater than 113 knots, or 209 kilometers per hour), and with it came rainfall so unprecedented that the National Weather Service issued a precipitation forecast stating that the impacts were unknown, beyond anything previously experienced. Houston was forecast to get more than 50 centimeters of rain. In fact, nearby Nederland, Texas, received an unprecedented 150 centimeters. All told, Harvey dumped 90 to 128 trillion liters of rain on the Houston area, so much rain that it depressed the Earth's crust by some 1.5 centimeters (as measured by GPS instrumentation).

The flooding from Harvey was catastrophic (figure 4.1). Rivers and bayous rose precipitously, overflowed, and submerged densely developed neighborhoods and major roadways. Intentional water releases from two reservoirs added to the disaster. Over three hundred thousand structures in the region were flooded and at least half a million cars immersed. Hundreds of thousands of people lost power during the storm. Forty thousand people evacu-

Figure 4.1. Catastrophic flooding in Texas during Hurricane Harvey. Courtesy US Air National Guard Staff Sgt. Daniel J. Martinez.

ated or took refuge in shelters, and thirty thousand water rescues were needed. Sixty-eight people died in Texas due to the direct effects of the storm, thirty-six in Harris County. The damage was estimated at $90 to $160 billion, second in cost only to Hurricane Katrina in 2005. Why did Hurricane Harvey intensify so rapidly and become a mega monster rainmaker?

When Hurricane Harvey passed over the Gulf of Mexico the ocean water was unusually warm and it ignited the storm. But the real combustion came when Harvey passed over a warm eddy: a circulating patch of water that was even warmer, several degrees warmer, than the surrounding ocean. Meteorologist Jeff Masters describes what happened: Harvey lingered over the warm ocean eddy for over six hours, and the extra energy it provided allowed Harvey's central pressure to fall a spectacular 15 millibars in just two hours. This combined with low wind shear and a moist atmosphere fueled Hurricane Harvey's rapid intensification.

When Harvey made landfall near Rockport, Texas, late on August 25, it was a major Category 4 hurricane. The pressure at its center was 937 millibars and sustained winds clocked in at 115 knots. Gusts were even stronger and destroyed homes and buildings. At the coast, pounding waves and a storm surge of up to 3 meters did additional damage. Storm surge is usually the most dangerous part of a hurricane making landfall, but in Harvey, it was the seemingly endless and epic rainfall.

Within twelve hours of landfall, Hurricane Harvey was downgraded to a tropical storm. But the worst was yet to come. Atmospheric high pressures lay to the northwest and southeast. These two systems fought to control and steer Harvey. But neither prevailed, and Tropical Storm Harvey stalled. The storm drifted east and southeast, lingering for days over Texas while winds from the south brought bands of rain born from the unusually warm waters of the Gulf of Mexico. Massive amounts of moisture were pumped into the atmosphere. The rainfall was torrential, especially to the north and east of the storm. So much rain fell that the flooded ground acted like the ocean, moisture rising from the water-soaked land, further fueling the storm. Over four days the storm lingered and an estimated 4 trillion liters of water were dumped on Harris County alone. Flooding in and around Houston was exacerbated by dense development on a floodplain and the paving over of the ground, which funneled water into rivers and low-lying areas. Harvey eventually meandered back into the Gulf of Mexico, and on August 30 made a final landfall near Cameron, Louisiana. The storm also spawned numerous, but fortunately brief, tornados in southeast Texas, Louisiana, Alabama, Mississippi, and Tennessee.

Harvey left devastation in its wake, but the 2017 hurricane season wasn't over. Following on the heels of Harvey were eight more hurricanes, including Irma and Maria. Both were also born from easterly waves coming off Africa, also went through periods of rapid intensification, and like Harvey, caused catastrophic destruction.

Irma was a beast (figure 4.2). In late August, while still far out in the eastern North Atlantic Ocean, Irma went from a tropical storm to a Category 3 hurricane in just thirty hours. Several days later, as the storm headed toward the Caribbean's Leeward Islands, it passed over an extremely warm ocean in a moist atmo-

Figure 4.2. Satellite (GOES 16/GOES East) image of Hurricane Irma heading for Florida. Courtesy CIRA/CSU and NOAA/NESDIS.

sphere. The result: Hurricane Irma became a Category 5 record-setter. With maximum sustained wind speeds of 160 knots, it was the strongest hurricane ever recorded in the open Atlantic Ocean. Irma headed west and to the surprise of experts maintained its Category 5 status for more than three days, with sustained winds of 160 knots for a record-setting thirty-seven hours.

At its peak intensity, Hurricane Irma struck and devastated Barbuda and then took aim at the US and British Virgin Islands. Some 95 percent of the buildings in Barbuda were destroyed, and conditions became so unlivable after the storm, the entire population of the island was evacuated to Antigua. The destruction was also epic in the British and US Virgin Islands, where deforestation was so severe that it could be seen from space. What had been lush green islands became desolate palates of brown.

But Irma wasn't finished. Though weakened, the storm swept through the southern Bahamas and approached the northern coast of Cuba. For days, Hurricane Irma tracked west, steered by a strong high-pressure ridge to the north. Forecasters predicted the steering ridge would weaken and the storm would turn north. But the various numerical models used to forecast the hurricane's track didn't agree as to when and where the turn would take place. Based on the forecast and potential for devastation, the governor of Florida declared a state of emergency five days before landfall.

On September 9, the high-pressure ridge weakened, and Hurricane Irma made its turn, heading northwest and straight for the Florida Keys. Before slamming into Cudjoe Key, the storm reintensified to a Category 4 hurricane with sustained winds of 112 knots. Devastation was again wrought. Increased wind shear and dry air then weakened the hurricane to a Category 3 storm before it struck Marco Island and headed north just inland of Florida's west coast. But Irma had blossomed in size, and now tropical storm–force winds extended out from the center for up to 640 kilometers. Damaging winds, heavy rains, and storm surge also struck the east coast of Florida. Jacksonville was hit with strong and persistent onshore winds, storm surge, and heavy rain resulting in some of the worst flooding in the city's history.

And then came Maria. On September 18, in just eighteen hours, Hurricane Maria went from a Category 1 storm to, like Irma, a Category 5 monster. The storm slammed into and destroyed much of the island of Dominica, then headed northwest toward the US Virgin Islands and Puerto Rico. Maria was a Category 4 hurricane with sustained winds of 134 knots when it struck Puerto Rico. The storm made landfall and moved west-northwest, churning its way across the island. Along with devastating winds, over a meter of rain fell, unleashing landslides and catastrophic flooding across the island. Based on buoy data, waves may have reached 6 meters high. The destruction was once again almost unimaginable. Thousands of people are estimated to have died in Puerto Rico due to the hurricane and its aftermath. Nearly all two million residents lost power. Infrastructure was so hard hit that tens of thousands of people left the island, and even nine months later many who stayed were still without power. Puerto Rico was forever changed by Hurricane Maria.

All told the hurricane season of 2017 was one of the worst in recorded history. Thousands of people were killed, and total costs have been estimated at upward of $300 billion. With regard to what made 2017 such a disastrous Hurricane season, several scientific questions loom large. What caused the hurricanes' rapid intensification, and can such changes be better forecasted? What role did climate change play in the extreme impacts of the storms? Were the hurricanes that occurred in such quick succession an example of clustering, and if so, what caused it? These questions and their answers relate to both our growing understanding and the lingering wish-we-knews when it comes to the Earth's most dangerous and powerful storms.

Hurricanes: The Known

As described in the accounts of Harvey, Irma, and Maria, the recipe for hurricanes can vary slightly, but the basic ingredients are always the same. A hurricane is born from some sort of weather disturbance, often described as a pulse of energy flowing through the atmosphere or an area of low pressure with disorganized storminess. Such disturbances may dissipate and move harmlessly on, or they may evolve into a more organized system, a strengthening storm. The conditions conducive to growth: a moist atmosphere, low wind shear, and warm ocean waters that are at least 26°C down to a depth of 18 meters or more. If the warmth does not extend deep enough, wind mixing will bring colder water to the surface and suck the energy out of a growing storm.

Winds begin to spin around a developing low-pressure system or storm due to the Earth's rotation and what is known as the Coriolis force. Hurricanes don't form on the equator because at zero degrees latitude this force is negligible. But just to the north or south of the equator, conditions are conducive to storm development, especially during warmer months. So most hurricanes form in the summer or early fall in two bands about the Earth, between about 4° and 30° north and south latitude (but predominantly in the north). Storms can form outside of these areas, but it is within this narrow band of heat that most of the sea's tempests are born.

As with Harvey, Irma, and Maria, many Atlantic storms begin with an easterly wave coming off Africa. During the summer, these atmospheric disturbances typically sweep off the continent every three to four days. Add the required atmospheric and ocean conditions, and easterly waves can grow into powerful hurricanes. The process by which this happens is fairly well understood. Air warmed by the underlying ocean rises, carrying moisture evaporated from the sea. As it rises, more air is sucked in from below and surface winds begin converging toward the area of low pressure created by the easterly wave. The converging air warms and picks up more moisture from the sea surface, and it too rises. High in the atmosphere, the rising moist, warm air cools and begins to condense. Clouds form and create burgeoning thunderstorms as vast amounts of latent heat energy are released. The rising air climbs higher. As more air converges and rises, storms become more organized and atmospheric pressure drops further. Winds begin to spiral around the central low pressure and an eye forms. When a storm's maximum sustained winds exceed 64 knots, it is considered a tropical cyclone—called a hurricane when it occurs in the Atlantic or northeast Pacific, a typhoon in the northwest Pacific.

The eye of a hurricane is roughly circular, typically between 10 and 100 kilometers in diameter, with relatively light winds. Recent observations, however, suggest that stronger than expected winds may occur in a hurricane's eye. It is sometimes so clear within the eye that, from above, the ocean is visible and, from below, blue sky can be seen. Surrounding the eye is the storm's eyewall, where the strongest winds and most intense rain bands occur. Intense hurricanes often strengthen when the eyewall goes through a replacement cycle.

Eyewall replacement cycles tend to occur over a period of eighteen to twenty-four hours. Rain bands in an outer ring of thunderstorms contract and rob the inner eyewall of its fuel—rising moisture and momentum. During the replacement process, the maximum winds decrease, atmospheric pressure rises, and a storm may appear less organized. But as a new eyewall forms, hurricanes may return to their previous strength or intensify even further. Why and exactly how eyewall replacement occurs is not fully understood. For forecasters, predicting and modeling it is a challenge.

The intensity of a hurricane is typically described on a scale going from a dangerous Category 1 to a catastrophic Category 5, based on the maximum sustained wind speed at the surface (peak one-minute wind at a height of 10 meters). Wind engineer Herb Saffir and meteorologist Bob Simpson designed the categories in the 1970s to effectively alert the public about potential storm damage. Early on, atmospheric pressure was included, as a proxy for wind speed. Modern aircraft reconnaissance flights now enable us to measure actual wind speeds. In 2012 atmospheric pressure and storm surge were officially removed from the category descriptions in what is now known as the Saffir-Simpson Hurricane Wind Scale. Storm surge was deleted because we now know that it depends on more than just wind speed. Bathymetry, the shape and nature of the coastline, the extent of the wind field, and the direction and speed of storm approach all influence where storm surge will occur and how high it will be.

Although helpful in describing a storm's intensity, the Saffir-Simpson categories do not necessarily reflect or correlate to the damage that can be done by storm surge, rainfall, or tornadoes, nor do they take into account the size of the storm or wind field. In light of recent monster storms, two questions have arisen. Should we continue to use a scale that doesn't include many aspects of a storm that can cause extreme damage? And should a Category 6 be added? Since the damage in Categories 4 and 5 is already described as "catastrophic," some people are not convinced another level is needed. What is beyond catastrophic—apocalyptic? After 2017, the people of Barbuda and Puerto Rico might say they've already experienced a Category 6 storm. Experts continue to explore how to improve warnings so that people will make the wisest possible decisions when faced with the impending impacts of an approaching hurricane.

One storm that wreaked havoc wasn't even a Category 1 hurricane or tropical cyclone. This tempest had devastating consequences and showcased the accuracy of forecasting while illustrating the challenges in communicating the dangers posed by powerful storms. This was Sandy—Superstorm Sandy.

SANDY, 2012. The storm was a late-hurricane-season classic. Sandy formed in late October in the Caribbean Sea and made its first landfall in Jamaica as a modest Category 1 hurricane. It later intensified and struck Cuba as a strong Category 3 hurricane, causing extensive and widespread damage. Coming off Cuba, Hurricane Sandy slowed and headed north, passed through the southern Bahamas, then gradually turned northwest and fell below hurricane strength. But as it interacted with the surrounding environment, Sandy swelled outward. Tropical storm–force winds now extended much farther—the wind field essentially doubled in size. Then, north of the Bahamas, Sandy reintensified. It was once again a hurricane, a really big hurricane. Reconnaissance aircraft flew into the storm to measure the winds. What they discovered was startling. The strongest winds were in the western side of the storm (based on the direction of movement, the strongest winds were expected in the eastern side), and the maximum sustained wind field was unusually large, with a radius of 185 kilometers.

Still several hundred kilometers offshore, Hurricane Sandy moved northeast, sweeping by North Carolina. A hint of an eye appeared in satellite imagery. Sandy was becoming more organized. Many hurricanes moving north or northeast off the US East Coast, especially in late October, move out over the open ocean. But on this day, October 29, the conditions were (at the time) unusual. To the north, an area of strong high pressure over Greenland blocked Sandy's track like an atmospheric wall. That meant the storm could turn either left (to the west) or right (to the east). To the west, a large trough in the jet stream pulled at the storm (figure 4.3). Hurricane Sandy hooked to the north and then northwest. Conditions also became more favorable for strength-

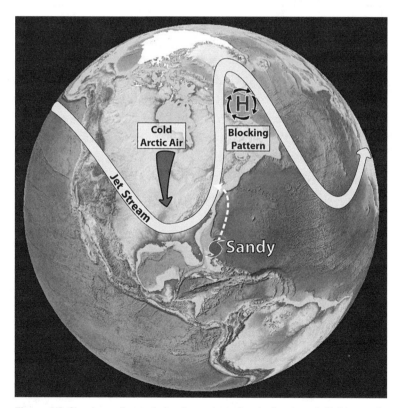

Figure 4.3. Steering patterns during Superstorm Sandy. Courtesy C. H. Greene et al., *Oceanography* 26, no. 1 (2013).

ening. Just over 370 kilometers off Atlantic City, New Jersey, Sandy's peak winds reached 85 knots.

Hurricane Sandy was spinning dangerously toward the coast with an extraordinarily large wind field. As it moved over cooler water and encountered cooler air, Sandy weakened and lost the warm core characteristic of a tropical storm. Officially, it was no longer a hurricane. It was a hybrid—a cross between a tropical cyclone and good old Nor'easter—called an extratropical or post-tropical storm. Sandy was one of the largest extratropical storms ever recorded in the Atlantic, and its winds gusted up to Category 1 hurricane strength. But the National Weather Service couldn't issue hurricane warnings, because Sandy was no longer officially a hurricane.

Days before Sandy was positioned to strike the New Jersey shore, weather forecasters knew they had a problem. Sandy was an enormous storm and would impact several states. It wasn't just the wind that would cause damage—given the size of the wind field and the direction of the storm's approach to shore, the storm surge was going to be bad, if not disastrous. They knew Sandy would lose its hurricane status before making landfall, though exactly when was unclear. Who, then, should issue the warnings and what should they be? Three days before the storm struck, the National Weather Service decided it would communicate the specific impacts in the landfall area with National Hurricane Center advisories. Local weather forecast offices would issue high wind watches and warnings. But what about storm surge?

Sandy made landfall just northeast of Atlantic City, near Brigantine, New Jersey, late on October 29. It was an extraordinarily large storm when it struck. Tropical storm–force winds extended out from its center for more than 740 kilometers. As expected, Sandy caused water levels to rise all along the coast, from Florida to Maine. When it made landfall, the states worst hit by storm surge were New Jersey, New York, and Connecticut. And it was a worse-case scenario, including the timing—Sandy hit at the same time as the full-moon high tide.

At the coast, powerful waves accompanied the storm surge. Homes and businesses were swept away or submerged. Along the

Figure 4.4. A roller coaster in Seaside Heights, New Jersey, submerged in the aftermath of Superstorm Sandy and severe beach erosion. Eric Thayer/Reuters.

New Jersey shore erosion was epic, some beaches losing up to 12 meters (figure 4.4). Fires ignited as gas mains ruptured. The highest recorded surge was at Kings Point, Long Island, where a tide gauge measured water levels more than 3.5 meters above normal. The storm surge also funneled up New York Bay and backed up the Hudson River. Lower Manhattan went dark as the ocean poured in. Seawater flowed through the streets, poured down subway entrances, and rushed into tunnels. In the United States, Sandy damaged or destroyed some 650,000 homes and caused seventy-two fatalities, forty-one due to storm surge. Eight and half million people lost power, and thousands were left homeless. The total cost of the storm is estimated to have exceeded $50 billion.

The forecast was right on the money, even six days out. But problems with communication led to confusion. During a storm update two days before Sandy struck, New York City mayor Michael Bloomberg announced, "Although we're expecting a large surge of water, it is not expected to be a tropical storm or hurricane-type surge. With this storm, we'll likely see a slow pileup of water rather than a sudden surge." He suggested Sandy's impact would be less dangerous than that of Hurricane Irene in 2011. Would people have prepared differently or evacuated if Sandy had been called a hurricane? That is an unfortunate unknown.

Since Sandy struck, two things have changed. When a post-tropical cyclone poses a significant threat to life and property, the National Hurricane Center can now continue to issue formal advisories, and hurricane watches and warnings can be kept in place. The National Oceanic and Atmospheric Administration (NOAA) has also made it a priority to provide explicit storm surge watches and warnings.

The dangers of and destruction caused by storm surge were widely seen and felt in Superstorm Sandy and illustrated that in storms it is not just the wind strength or "Category" that matters. The approach to the coast and size of the storm can, in some cases, be equally if not more important than the wind speeds. The evolution of Superstorm Sandy also highlighted the importance of understanding and predicting how a storm interacts with the surrounding ocean and atmospheric environment.

Hurricanes: The Unknowns

Forecasting and Advances in Technology

In 1995 Michigan congressman Dick Chrysler stood up in the once-hallowed halls of the Capitol in Washington, DC, and asked why we need the National Weather Service when he could simply turn on the Weather Channel to get the forecast. The science and meteorological community were aghast at the man's sheer and blatant ignorance. StormCenter Communications meteorologist Dave Jones put it this way: It's kind of like asking why we need cows when there's plenty of milk in the stores.

People can indeed turn to their television, radio, computer, or mobile device and get up-to-date forecasts or weather alerts

24/7. Weather and hurricane forecasts are more accurate and accessible than ever before in the history of human civilization. But they don't just fall from the sky; they are based on data, which are collected, processed, analyzed, and then integrated into sophisticated numerical models by people at the National Weather Service and its partners. Numerical models generate visualizations, synoptic overviews, and forecasts, which along with expert interpretations are then given to local and private providers—like the Weather Channel.

Much of the improvement in hurricane and weather forecasts comes from advances in technology and access to better, more comprehensive data. A ginormous amount of data is needed to accurately forecast where a hurricane is headed, its changing intensity, rainfall, winds, and storm surge. Billions of data points are drawn from satellites, weather balloons, airplanes, buoys, and ground observing stations around the world each day.

Not only is the amount of data needed enormous, but it must also be rapidly processed and integrated into complex numerical models. And not just one model. Multiple models are used in weather and hurricane forecasting to account for their differing strengths and weaknesses. And they're run using varying sets of initial conditions and changing atmospheric and ocean conditions. This also takes a tremendous amount of computing power.

In January 2018, NOAA announced a major upgrade to its supercomputing capability. The upgrade is especially timely, as it comes at the same time as a wealth of new data from NOAA's latest satellites, GOES-East, NOAA-20, and GOES-West. With spectacularly improved resolution and less time between images, the new satellite data has garnered great excitement in the meteorological community. Researchers and forecasters are now able to see things they've never before witnessed—like cloudbursts from thunderstorms in hurricanes and eyewall replacement cycles. The additional supercomputing power is also vital as NOAA implements the next generation of global forecast models. But even with today's highly complex numerical models, some things, which seem rather simple, remain challenging to capture—such as clouds. Clouds are often too small to be resolved in model grids. Aerosols are another factor that remains difficult to integrate into weather and storm simulations.

Along with the invaluable imagery obtained from earth-orbiting satellites, hurricane forecasting has radically improved due to direct wind and atmospheric pressure measurements provided by air force reconnaissance flights. Ironically, it was an audacious bet that led to the first purposeful flight into a hurricane.

It was July 1943, and across the breakfast table at the Army Air Forces Instructors School in Columbus, Texas, a wager was made. Veteran British pilots were needling the base commander, Lt. Col. Joseph Duckworth, about the frailty of US planes. A hurricane was approaching Galveston, and American airplanes were being flown away from the storm's path. Reportedly, Duckworth, fed up with the ribbing, bet the British pilots a cocktail that he could fly one of the airplanes into the storm and return safely. The British accepted the wager. Duckworth and Lt. Ralph O'Hair fueled up an AT-6 Texan single-engine trainer and took off from Bryan Field. As they approached Galveston, the people manning Houston's air traffic control tower reportedly radioed to ask if the pilots realized they were headed into a hurricane. Hearing of their plans, the controllers responded with a simple request: tell us where to look for the wreckage.

Fortunately, the flight was successful. Upon its return to the airfield, the base weather officer met the airplane, jumped aboard, and Duckworth flew back into the storm to collect data. Since then, the observations and data collected by the Air Force Hurricane Hunters have become a critical part of providing reliable hurricane forecasts. Still, it's risky business and not for the lighthearted. NOAA meteorologist and hurricane hunter Rob Rogers remembers some of his more frightening experiences: planes hit by lightning, three engines lost during flight, and setting off for what his team thought would be a Category 1 storm, only to learn on arrival that it had evolved into a scary Category 4 hurricane.

What else is trending in meteorological technology today? Drones. Remotely operated or autonomous vehicles can be programmed to fly into dangerous conditions or hard-to-access locations, and operate over long periods. In 2017 researchers aboard NOAA's P-3 Orion flying into Hurricane Maria deployed Coyotes, the latest and smallest hurricane hunting drones. The Coyote, with a wingspan of 1.5 meters and weighing about 6 kilograms, was originally designed for military purposes. Dropped from an airplane with a small parachute, its winds and rudder extend. It is piloted from the deploying airplane and can fly for up to an hour before dropping into the sea. The drones are particularly useful in conditions that are too turbulent and violent for crewed aircraft, such as at low altitudes in or near a hurricane's eyewall. Seven Coyotes were dropped into Hurricane Maria to determine the radius, velocity, and exact location of the storm's maximum winds. Before being lost, one drone was able to fly for about forty-two minutes down to 122 meters. Two other Coyotes deployed in Maria's eyewall descended to 335 meters before being lost. Still in the experimental stage, Coyotes hold great promise for data collection in hurricanes. Hopefully more will soon be flying into the unfriendly skies.

Another unmanned aircraft being deployed to learn more about hurricanes is NASA's Global Hawk. Acquired originally from the US Air Force, the thirteen-meter-long drone has a wingspan of over 35 meters and is powered by a Rolls-Royce AE3007H turbofan engine. The unmanned aircraft can fly for up to twenty-four hours, with a range of some 15,700 kilometers, and has a communications system that allows for worldwide operations. Two pilots are required to fly the Global Hawk, but they can be far away sitting comfortably in a parked trailer. The Global Hawk is flown above hurricanes and was used in Cristobal (2014) and Gaston and Matthew (2016).

In 2017 the Global Hawk became part of a joint NASA-NOAA project to better understand the rapid intensification of hurricanes in the Northern Hemisphere. Planned flights in the program will be twenty-four hours long at an altitude of 18 kilometers. Instruments aboard or deployed from the Global Hawk will measure wind velocity, pressure, temperature, humidity, and cloud moisture content and assess the overall structure of a storm system. A new Doppler radar system aboard the drone will also allow for observations of a storm's vertical velocity profile. It's hoped that this joint-agency project will improve our understanding of rapid intensification, which remains the most challenging aspect of a hurricane to forecast—a key wish-we-knew topic for most meteorologists. When a storm nears the coast, it is also the cause of many an expert's nightmares.

Rapid Intensification

In 2016 Hurricane Matthew grew from a tropical storm to a Category 5 hurricane in just thirty-six hours. Scientists were stunned. Moderate to strong vertical wind shear in the area had suggested strengthening was unlikely at best. None of the computer models predicted the storm would significantly intensify. And yet—Matthew grew rapidly into a monster storm with 143-knot winds. The hurricane would wreak havoc in Haiti, cause damage in Florida, and generate severe flooding in South and North Carolina. Why Hurricane Matthew rapidly intensified in conditions thought to be unfavorable remains a puzzling science unknown.

Rapid intensification is defined as when a hurricane's maximum sustained surface winds increase by 30 knots or more within a twenty-four-hour period. Being able to forecast this has been called the holy grail of intensity forecasting. The dynamics of rapid intensification remain poorly understood and are hard to predict. This is undoubtedly because the processes involved are highly complex, range widely in scale, and are affected by both internal and external forcing factors. Internal factors include the behavior of clusters of intense thunderstorms known as convective bursts, the tilt of the storm's vertical structure, changes in atmospheric pressure, and eyewall replacement cycles. External influences include wind shear, upper ocean temperature, atmospheric moisture, and interaction with other weather systems or longer-term cycles in the coupled atmosphere-ocean system. Simply put—it's really complicated.

Satellite-based sea-surface temperature data has helped uncover one now-notorious cause of rapid intensification in hurricanes. During the summer and early fall, warm ocean water flowing from the Caribbean Sea into the Gulf of Mexico forms a looped current (not surprisingly known as the Loop Current). Sometimes, swirling patches of seawater form within the Loop Current and break off. Known as warm core eddies, they can contain seawater that is several degrees warmer than the surrounding Gulf. If a hurricane passes over a warm core eddy, it can provide the superfuel that drives rapid intensification. This has happened in numerous storms, including Opal (in 1995), Katrina (2005), and Harvey.

Another potential source of hurricane superfuel lays just off the southeast US coast—the Gulf Stream. Here, warm waters originating in the tropics stream and meander northward. Within the Gulf Stream, seawater temperatures can be significantly higher than in the surrounding ocean. If a hurricane moves over the Gulf Stream, it also has the potential to rapidly intensify, as happened in Hugo (1989), Andrew (1992), and Irma. But—and here is another big but—if the warm water in the Gulf Stream or an eddy is too shallow, a passing hurricane will not rapidly intensify. Unfortunately, satellite sensors detect temperature only at the very surface of the sea, and such data alone can indicate only the potential for rapid intensification, not a sure thing. Once again, adding to the challenge of intensity forecasting.

Scientists hope that with higher-resolution computer models that couple the ocean and atmosphere, NOAA's newest satellites, and data from subsurface oceangoing buoys and unmanned aircraft, such as the Global Hawk, our understanding of and ability to forecast rapid intensification will improve. This is particularly important for storms approaching the coast. Rapidly intensifying hurricanes close to shore put populations at great risk and can leave little time for adequate preparation or evacuations.

The reliability of hurricane track forecasts has steadily and greatly improved over time, especially within a few days of landfall. Intensity forecasts have not improved equally. Scientists and forecasters are working hard to better understand rapid intensification and to better forecast it. They want to know not only which storms will strengthen abruptly, but also why some that they think should don't.

Climate Change

There is no question or credible debate: climate change is occurring. Climate change is not the cause of hurricanes, but it is worsening their impacts, as exemplified by the catastrophic flooding during Harvey. Some of the factors and processes at work are already well recognized. As the ocean warms, more heat and moisture—fuel for hurricanes—become available. Changes in the jet stream and wind patterns can cause hurricanes to linger longer, exacerbating flooding and damage. And rising sea level increases the impacts and inundation from storm surge. Data suggest that in a warming climate more hurricanes may also rapidly intensify. How else will climate change influence hurricanes and their impacts? Will there be more hurricanes or just more intense hurricanes? Will the tracks hurricanes take deviate significantly from those of the past?

A growing consensus suggests that the number of hurricanes per year probably will not increase. But with higher ocean heat content available to fuel storms, the number of intense hurricanes may rise, as may the strength of the storms. It is less clear if typical hurricane tracks will shift or alter. However, as warmth spreads north and steering systems like the jet stream change, hurricanes may move in less expected ways, as in Superstorm Sandy's westward turn toward shore. With improved numerical models and increased data collection, scientists hope to better understand and forecast hurricanes in the context of climate change. One phenomenon that plays a role in this is El Niño.

El Niños and Hurricanes

El Niño, sometimes called the El Niño Southern Oscillation (ENSO), is a coupled ocean-atmosphere phenomenon in the equatorial Pacific that, based on past history, tends to occur about every four to seven years. During El Niños, the trade winds relax and a warm pool of ocean water flows from west to east across the Pacific. Drier conditions in the western Pacific cause drought and wildfires in locations such as Australia, while warmer, wetter conditions in the east typically result in heavy rains in California and Peru. It is unclear if El Niños will become stronger or more frequent in a warming world. But because of their influence on storms, especially in the Atlantic, El Niños add another unknown to the equation for future hurricanes.

During El Niño years, wind shear over the Atlantic increases and appears to prevent or reduce the formation of hurricanes. Tropical scientist Phil Klotzback at Colorado State University, who works with a team of researchers to predict the number and strength of hurricanes each year, says forecasts would be greatly improved if they could predict El Niños sooner and better integrate them into their calculations.

Sometimes though, El Niños don't behave as expected. In 2016 the central north Pacific became unexpectedly warm due to an El Niño. The warmth had been expected to move farther east. Was this normal for an El Niño? Are our records too short to reveal long-term patterns and variations in El Niños?

Storm Surge

Moving water is a powerful force. Even half a meter of water running across a road can sweep a person off their feet. A little more and cars begin bobbing. One cubic meter of water weighs 1,000 kilograms—about the weight of a Smart car. Imagine being slammed into by a Smart car in the midst of a hurricane. That's called storm surge. Add waves, wind, and tides and you get an extremely powerful force that is responsible for most hurricane-related deaths and that can cause catastrophic destruction.

In hurricanes, wind and flying debris are enormously dangerous, but storm surge is the most frequent killer. In Superstorm Sandy (2012), forty-one people were killed by storm surge. In Hurricane Katrina (2005) some eighteen hundred people lost their lives, most of them to storm surge. During Katrina, the rush and rise of the sea reached up to 8 meters above normal tide levels and ravaged the Mississippi coast. Seawater rushed inland for kilometers. In Louisiana, levees were overtopped and breached, leaving much of New Orleans and its suburbs underwater. The disaster was one of the worst and costliest in US history and put a tragic spotlight on the dire need for better hurricane preparation and response, especially when faced with the needs of the elderly, sick, and poor.

Storm surge is not an unknown. We know what creates it and to some extent can forecast when and where it is likely to occur. But getting it exactly right and forecasting how much surge will occur is another complicated challenge.

Storm surge is described as the abnormal rise of the ocean generated by a storm. It occurs mainly due to wind and waves pushing seawater ashore faster than it can drain away. The low atmospheric pressure that accompanies a hurricane also acts to suck the ocean slightly upward. When storm surge is accompanied by waves or occurs at high tide, the impacts are even worse. Other factors that create and influence storm surge include a storm's intensity, speed, and direction of approach to shore, the size of the wind field, offshore bathymetry, and the nature of the shoreline. No one variable alone can be used to predict or model storm surge.

Today, sophisticated computer models can simulate historical, predicted, or hypothetical storm surges. But they don't include waves, rainfall, river flow, or the timing of tides. Scientists and forecasters are working hard to collect more data and improve storm surge models. They are also exploring ways to better communicate the extreme danger posed when the sea comes ashore in a storm. Many people still don't realize the power of storm surge or understand that forecasts rarely present worse-case scenarios.

In years past, storm surge forecasts were made relative to mean low water, high tide, or sea level. But few people know offhand where those levels are or how they relate to their house or parking space. They need context: how far inland will the water come, and how high relative to something familiar? To make storm surge warnings more meaningful to more people, forecasters now provide potential surge levels relative to "normally dry ground." But with climate change, what's normally dry is changing. Rising sea level may make what was normally dry ground not so normally dry. Storm surge will be worse in the future, but how much so remains a wish-we-knew.

JOAQUIN, 2015. From its origin and throughout its growth, Hurricane Joaquin vexed those who sought to predict its path, and it highlighted conditions that make forecasting especially difficult. The storm began in an unusual location, as an atmospheric disturbance in the eastern North Atlantic off the Canary Islands—not where most hurricanes are born. It moved west, then southwest, and finally south toward the central Bahamas, becoming Hurricane Joaquin on September 30, some 314 kilometers east-northeast of the island San Salvador. With sea-surface temperatures warmer than normal, Joaquin grew quickly into a Category 4 beast. The storm's winds were sustained at 120 knots as it moved slowly to the southwest.

Meanwhile, over the US East Coast an upper-level low pressure or trough was deepening; this caused Hurricane Joaquin to slow even further. The storm then took an unexpectedly sharp turn to the south-southwest. Where would Joaquin go next? The answer was unclear.

"To accurately understand what the weather will do days in advance," meteorologist Eric Holthaus writes, "you have to observe it perfectly—and perfectly understand the underlying physics. We actually do the second part better than the first—there is just no way to launch enough weather balloons or satellites to monitor the entire Earth system, down to the millimeter."

As it turned out, Hurricane Joaquin wasn't going anywhere; for two long days, the storm lingered in the Bahamas. And that's where the *El Faro*, a cargo ship in poor condition, ran bow-first into the stalled storm. On October 1, the ship was reported dead in the water and listing. A distress signal followed, but hurricane conditions hampered search and rescue efforts. Tragically, all thirty-three crewmembers were lost when the ship sank in more than 4,500 meters of water.

With the surrounding conditions and steering patterns evolving rapidly, Hurricane Joaquin's next move remained a puzzle.

Would the storm head north, or turn northwest and make a US East Coast landfall? The numerical models being used to forecast the hurricane's path disagreed wildly. It's not that the models were bad or inaccurate. The environment surrounding the storm was changing so rapidly, it was hard for the models to integrate new conditions fast enough. Eventually, as more data came in from aircraft and observing stations, the models came into agreement and an accurate forecast was made. Hurricane Joaquin moved northeast, away from the Bahamas and out to sea. The storm highlighted the difficulties of forecasting in rapidly changing conditions and the need for continually updated observations. Sometimes a lack of movement or just a tiny wiggle or wobble in a hurricane's track can have disastrous consequences.

CHARLEY, 2004. On August 13, 2004, Hurricane Charley moved off Cuba, rapidly intensified, then barreled north with 150-knot winds. The Category 4 hurricane skirted Key West and turned slightly to the northeast, heading toward Tampa Bay almost parallel to shore. For four days most of Florida's southwest coast was in what was then called the cone of uncertainty—the area most likely to be hit based on the track and previous storms. Hurricane watches and warnings were in affect all along the coast. Then, six hours before the expected landfall—a wobble occurred.

Hurricane Charley's eye jigged slightly to the east. Tampa Bay was no longer the probable landfall. Based on radar imagery and data from the Hurricane Hunters, the National Hurricane Center adjusted the forecasted track to show the storm's eye moving onshore well south of Tampa, near Charlotte Harbor, as a Category 3 hurricane. It was 1 p.m. Hurricane Charley made landfall a little over two hours later. At least ten people died due directly to the storm. In Charlotte County, some 80 percent of the buildings were destroyed and a storm surge of more than 2 meters inundated coastal Lee County. Numerous schools in the region were heavily damaged, and light poles along the highways were bent over like straws. Punta Gorda took the full brunt of the storm and was essentially leveled.

Charley was strong and fast-moving, but a relatively small

storm—its eye was less than 10 kilometers wide when it made landfall. Yet it was deadly and destructive, and brought with it important lessons. When hurricanes travel parallel and close to shore, a small wobble can quickly change where they strike. And although expected landfalls are typically stated as a specific point, in truth even small storms impact broad areas—not just one dot on a map.

After Hurricane Charley, many people complained about the forecast. But the forecast had been accurate—the area of landfall had been in the area at risk for days. The problem was how people perceived and responded to the forecast. Many had focused too much on the central track line, which for days had targeted the Tampa Bay area. When the hurricane wobbled and struck farther south, they were unprepared. There was great debate after the storm about whether forecasts should even show the central track line. Today, graphics allow users to hide the track line and just show the cone of uncertainty—now referred to as the forecast track cone. Additional efforts were and are being made to better educate alert providers and the public about forecasts and to update them more frequently. NOAA's new satellite imagery and data should markedly improve these efforts.

Hurricanes regularly wobble along their tracks. Can these wiggles and jigs be predicted? In the short term, if radar and observations are available—probably. When the shifts are forced by large-scale systems—maybe. When wobbles occur due to the chaotic nature of a storm—probably not, especially not more than a few hours in advance.

Along with highlighting a hurricane's potential to wobble and the consequences, Hurricane Charley also brought into focus an issue that has long frustrated forecasters: how to most effectively warn people about an impending storm and its dangers. Meteorologist Max Mayfield, a former director of the National Hurricane Center, is among those who wishes he knew how to motivate people to respond appropriately to a hurricane threat. He notes that for many storms, including Katrina and Sandy, the forecasts have been pretty good—as good as the science allows them to be—and yet there is still an unacceptable loss of life. When hurricanes approach land, efforts are now made to emphasize and provide more information about the potential impacts of wind, precipitation, and storm surge, as well as tornadoes. But what is the best way to show not just the intensity of a storm but also the size of the wind field and the breadth of the regions at risk? And how, when forecasts are never 100 percent certain, do we get people to make preparations or even to evacuate? Researchers, forecasters, and social scientists are working to come up with the answers.

Tornadoes

As hurricanes make landfall, they bring with them the ingredients for an unwanted and destructive companion—tornadoes. A tornado is a violently spinning column of air extending from the base of a thunderstorm to the ground. The ingredients needed for a tornado and provided by hurricanes include supercells, or rotating, well-organized severe thunderstorms, and rising warm moist air. As a hurricane moves onshore, friction from the land slows the flow of air in its lower levels; this creates another factor needed to produce a tornado—vertical wind shear.

Tornadoes can occur during landfall or over the next several days as a hurricane moves inland. They typically form in a hurricane's spiraling bands of rain but have also been observed in the eyewall. Research suggests that tornadoes may be most likely to form in the right-front quadrant relative to a hurricane's direction of movement. But the association is not clear.

Scientists today understand the conditions that generate tornadoes, but the exact processes by which they form are not well understood—in other words, it is still a big, spiraling wish-we-knew. Advances in technology, particularly radar, allow forecasters to identify conditions such as supercells and rotation that can spark tornadoes. But radar cannot access the area at the very base of a storm—which is exactly where tornadoes form. Unmanned aircraft or drones are one of the latest tools being used to record conditions prior to and during tornadoes. The hope is to forecast tornadoes sooner and provide warnings further in advance.

Confirmation that a tornado has actually formed is also problematic. Currently, it comes from direct observation by people or cameras. It is hoped that in the future improved technology will provide a means to confirm tornadoes without putting observers at risk. The official confirmation that a tornado has struck, and of its strength, comes after the fact through an assessment of the damage done.

Hurricane and tornado science will continue to evolve as we seek to better understand and forecast Earth's most dangerous storms. Hurricanes will continue to take oddball tracks, wobble, and rapidly intensify, just as we will continue to try to better forecast them. With rapidly advancing technology we are learning more every day.

Today, forecasts are considered very reliable three days out, but it is hoped in the future this will be extended to five or more days. Still there will always be at least some uncertainty in forecasts, and difficult decisions will still have to be made in advance. For local, regional, or national authorities, those decisions depend on understanding the threats and determining how much risk is acceptable. For individuals, it is often a personal choice, how much to prepare for a hurricane and whether or not to evacuate. Again, education and understanding of the threats are critical. Going to the beach to watch or even to surf as a hurricane approaches, even a Category 1 storm, exemplifies a sad lack of understanding, poor decision making, or plain old stupidity. So does driving through floods or surging seawater. Reckless behavior in storms also puts first responders at risk. Preparation is clearly key, though sometimes even in shopping for supplies, people make surprising choices: during 2011's Hurricane Irene, one large supply and grocery store reported that the most popular item purchased during preparations was strawberry Pop-Tarts. Then again, they probably never go bad and can be eaten even if the power goes out.

Again, there is strong consensus among scientists and forecasters: we need to continue to invest in scientific research and in learning more about hurricanes, but we must also do more to prepare and educate everyone, and to respond effectively when danger is imminent. This is increasingly important in a warming world in which extreme weather and its impacts are on the rise.

For more on the basics of hurricane science, storm history, and forecasting, see the list of readings and references at the back of this book.

5

Rogue Waves, Landslides, Rip Currents, Sinkholes, and Sharks

Rogue Waves

NIGHTTIME, CROSSING THE ATLANTIC OCEAN. As a full moon rose over the glassy calm sea, the revelry aboard the ship grew raucous. It was an epic celebration—New Year's Eve on the high seas aboard a luxury cruise ship. Officers on the bridge passed on the champagne but were afforded a spectacular view of moonlight shimmering atop the ocean. Just before midnight, the captain felt a strange vibration. "Something's off," he muttered. He looked to the officer at the radar and grabbed a pair of binoculars. Raising them to his eyes, he gasped in disbelief.

Racing toward the ship was a wave of enormous proportions, so high that as it neared its crest blotted out the moon. The captain shouted to the helmsman to turn hard to starboard and put the starboard engines full astern. They had to turn into the wave to avoid being hit broadside. The ship slowly started to turn. The watery colossus drew closer. The captain shouted to use the bow thrusters to turn the ship faster and to sound the alarm.

As it bore down on the cruise ship, the 45-meter-high wave began to break. The ship was still nearly broadside to the wave. It was going to be hit in the worst possible position. The ship started to list. Whitewater smashed through the windows of the bridge. The cruise ship began to roll until finally it went fully over—capsized. Decks turned to ceilings and revelers went flying. No one could have predicted the disaster—it was a rogue wave.

Fortunately, it was also just a movie—*Poseidon*, the 2006 remake of the 1972 classic *The Poseidon Adventure*. The movie's special effects were fantastic, particularly the approach of the towering wave in the moonlight—but luckily, maybe not so realistic. With advances in radar, ship construction, and propulsion, once a rogue wave is detected, it is hoped that most cruise ships will have enough time and power to maneuver into it and minimize damage. In 1998 the captain of the *Queen Elizabeth II* did just that: detected a rogue wave on radar and turned into it. When a 27-meter-high wall of water hit, the ship suffered little damage. In 2005, a 21-meter rogue wave struck the nearly 300-meter-long cruise liner *Norwegian Dawn* off the coast of Georgia. Windows were smashed, deck chairs went flying, and four passengers were injured—but again, damage was relatively minor. As for *Poseidon*'s giant wave: it is unlikely that a rogue wave would occur in a flat, calm sea, and although its 45-meter height is theoretically possible, the largest rogue wave documented so far was a bit less—36 meters high.

Rogue waves are essentially waves that are significantly higher than those in the surrounding ocean—sometimes defined as greater than two times the significant wave height (the average of the highest one-third of waves over a given period). Once thought mythical or incredibly rare, rogue waves are now known to occur more frequently than previously believed. Still, there remain some tall wish-we-knews when it comes to the sea's freakishly big waves.

Rogue waves have now been observed from ships and oil platforms, documented by instruments on buoys, and detected by radar aboard earth-orbiting satellites. These observations have helped reveal how often rogue waves form and where they are most likely to occur. For instance, in 2004 instruments moored on the seafloor in the Gulf of Mexico about 160 kilometers south of Mobile Bay, Alabama, documented a 32-meter wave spawned by Hurricane Ivan. During the European Union's MAXWAVE project, scientists analyzed some thirty thousand radar images looking for rogue waves. During the three-week study period they detected ten rogue waves, each more than 23 meters high. In another study, scientists using data from two buoys in the North Pacific analyzed two million wave groups and discovered more than three hundred rogue waves. Clearly, rogue waves are not mythical, nor do they occur only once in thousands of years as previously thought.

Still, by nature, rogue waves are relatively infrequent and transient; most don't last long. Two vexing questions are how exactly do they form, and can they be forecasted in advance to warn oceangoing vessels?

Rogue waves appear to form in several ways, though the exact processes involved are poorly understood. Just off the coast of South Africa is one the most treacherous areas of the sea. Each year, tankers transiting the region are struck by giant waves and seriously damaged. It is prime birthing ground for rogue waves because here, easterly driven wind waves run into the westerly flowing Agulhas Current. We now know that rogue waves more commonly form in such areas, where wind-driven waves interact with powerful ocean currents flowing in the opposite direction. This happens as well in the Kuroshio Current off Japan and the Gulf Stream off the east coast of the United States.

Storms, such as Hurricane Ivan, can also spawn freakishly large waves, as can wave trains traveling at different speeds that by chance merge and create towering, though short-lived, waves. Scientists are working to identify specific locations around the world where rogue waves are most likely to form, such as off South Africa, and to find ways to create reliable forecasts for them. As of yet, however, we cannot predict where and exactly when a rogue wave will form.

What is the maximum height for a rogue wave? Calculations suggest that theoretically a rogue wave could grow to be 60 meters high—that's taller than the Statue of Liberty. But given that the physics are not well understood, it's hard to say if that is truly possible.

Throughout history, rogue waves have damaged and sunk numerous oceangoing vessels, and more extreme waves will undoubtedly occur in the future. The big and towering questions are will ships be prepared, can we reduce the risk through improved understanding, and might advance warnings one day be a reality?

Landslides

OSO, WASHINGTON, 2014. The morning of March 22 was clear and sunny in the mountains of the Northern Cascades. Fifty homes sat peacefully at the base of a hillslope in Oso, a town about an hour's drive northeast of Seattle. Then at 10:37 a.m., meters of mud mixed with boulders, trees, and other debris plowed through the town, burying roadways and destroying forty-three homes. Forty-two people died, and ten were seriously injured. In total, the landslide moved 18 million tons of sand and mud, enough to cover some six hundred football fields more than 3 meters deep. Economic losses were estimated at $50 million. It was one of the worst landslides in US history.

A county emergency manager would later claim the landslide came out of nowhere, an unforeseen event in a very safe area. In fact, however, a 1999 report to the US Army Corps of Engineers by geomorphologist Daniel Miller and his wife, Lynne Rodgers Miller, had warned that the slope above Oso had "the potential for a large catastrophic failure." The geology of the area indicated it was prone to landslides with documented events in 1949 and 1951 (another would hit in 2006). Evidence suggested that landslides had also occurred in the more distant past. Were those living in the area or the authorities responsible for zoning and development not told of the report, or was it ignored?

An extensive investigation after the landslide identified several factors as contributing to the destablization and mobilization of the hillside: the preceding three weeks of intense rainfall; alterations of the local drainage and groundwater recharge due to previous landslides; changes in the stress and stability of the slope, again due to prior events; and possibly the role of logging on the upper slopes. As with many of the other phenomena described in this book, we can now recognize where landslides are likely to occur, but we still cannot say when they will happen, how big they will be, or exactly where they will go. And clearly, communicating the dangers posed by landslides remains a significant challenge.

Landslide or UFO

The pattern in the image is curious (figure 5.1). It appears that an otherworldly object hit the frozen hillside, disrupting the snow and ice, then rolled downward and slid out over an icy plain at high velocity, creating a narrow channel in the snow hundreds of meters long—at the end of which sits a strangely linear, fifty-

Figure 5.1. Rare runout landslide on frozen landscape of South Georgia Island. Courtesy Google Earth and Dave Petley (Landslide Blog).

meter-long snow-covered object. At least that's what the people at "Secureteam10" would like you to believe. Its video purporting a not-from-this-planet impact scenario has been viewed hundreds of thousands of times. Sorry to disappoint conspiracy theorists, but this is not the covered-up crash site of Kal-El, a.k.a. Superman, or of any other alien on an Earth fly-by. It is what is known as a runout landslide, on the island of South Georgia, north of Antarctica.

Landslides can take numerous forms, such as rock falls, avalanches, or mudflows. In some cases, isolated boulders or blocks can travel surprisingly long distances as runouts. Dave Petley, who writes an excellent blog on landslides, revealed the non-UFO truth about the icy landslide in South Georgia. While the long narrow runout is unusual and a poorly understood phenomenon, he explains, the low-friction slipperiness of the snow and ice probably contributed to its occurrence.

Landslides happen all across the world and are much more common than most people think. Each year they kill thousands of people and cause billions of dollars in damage. In the United States alone, landslides are estimated annually to cause up to fifty fatalities and cost about a billion dollars. They can happen literally anywhere—mountains, grassy plains, forests, underwater—and are sometimes associated with volcanoes or earthquakes. Dave Petley is particularly curious about landslides that happen on slopes during earthquakes. It's a complex and dynamic situation, he explains, and the answers may vary from case to case. It's also difficult to know when, where, and how to instrument a slope adequately to capture a major earthquake.

Landslides—essentially the gravity-driven downslope movement of rock, debris, soil, or earth—include mudflows, debris flows, volcanic lahars, and rock falls. Once set in motion, material in a landslide may flow, slide, spread, fall, topple, creep, or roll. The type of landslide that occurs depends on local and regional conditions, slope, and rock type. Weather and climate play a role as well, as can human activities such as logging, other removal of vegetation, undercutting of slopes, or changing drainage patterns. And as people in California know all too well, wildfires can make an area much more susceptible to landslides.

MONTECITO, CALIFORNIA, 2018. In the early morning hours of January 9, a deep rumbling was heard by residents living just southeast of Santa Barbara in Montecito. Soon afterward, a swiftly flowing wall of mud, rock, water, and debris swept through and literally buried much of the community. The consistency of wet cement, it rushed through streets, carried away cars, and demolished homes. Riding atop the mud and debris were giant boulders that bulldozed whatever lay in their path (figure 5.2). More than twenty people died, and hundreds of homes and buildings were destroyed or badly damaged. While scientists and emergency managers had warned that a mudflow could occur, they could not predict exactly when it would happen or where or how far it would go.

The conditions that triggered the Montecito mudflow actually

Figure 5.2. Montecito, California, in the aftermath of 2018 mudflow. Kyle Grillot/ Reuters.

began in December 2017 when one of the state's largest wildfires, the Thomas Fire, burned more than 280,000 acres across Southern California. The wildfire burned away vegetation that provided a protective canopy and whose roots helped bind the soil together. The fire also destroyed the covering base layer of ground litter. The low-lying shrubland that burned, known as chaparral, had leaves with a waxy resin coating that helped the plants retain moisture and resist drought. When the leaves burned, the waxy substance volatilized and then settled, creating a water-resistant coating over the ash and soil.

Then came the rain. Weeks after the fire, the Montecito area was hit was torrential rains. Water poured from the sky, and much of it ran off the now waterproofed soil and ash. As the runoff moved downslope, it grew in volume and momentum, picking up sand, rocks, debris, and even giant boulders—creating a massive and deadly mudflow. But it wasn't the area's first such landslide. Montecito was built on old mudflows.

In the aftermath of the disaster, scientists mapped and studied the Montecito mudflow. The information gained will be used to better predict future events and to create improved hazard maps and warnings. Unfortunately, the area remains at risk for future mudflows.

Intense rainfall is one type of landslide trigger—another is earthquakes. Perhaps nowhere else on Earth has this been more tragically showcased than in Wenchuan, China, where, in 2008, some eighty thousand people lost their lives in a catastrophic 7.9 magnitude earthquake. It is estimated that as many as a quarter of those killed were caught in landslides that buried large sections of towns. With thousands of landslides triggered, the event also caused mass quantities of sediment to slide into and block rivers, leading to subsequent floods. While the character of the debris flows has changed with time, the region continues to be plagued by landslides as happened in 2010, 2011, 2013, and 2016. The earthquake-triggered landslides and subsequent river impacts around Wenchuan left the scientific community reeling and highlighted a lack of understanding about such events along with some big and dangerous wish-we-knews.

As scientists strive to better understand landslides and their causes, technological innovations are helping. New tools such as satellite imaging, radar, precise laser mapping, and drones are allowing scientists to identify and map landslides as never before. Numerical and laboratory models are being used to simulate how landslides happen and determine where they are most likely to occur. Combining these technologies, scientists, emergency managers, and others are also developing warning systems that will potentially be able to "nowcast" landslides on a regional and global basis.

In Southern California, the National Weather Service and the US Geological Survey are teaming up, using rainfall measurements and estimates to generate flash flood and debris flow early-warnings for recently burned areas. Another system, called the Landslide Hazard Assessment Model for Situational Awareness or LHASA, is doing something similar on a global scale, combining satellite-based precipitation measurements with sophisticated computer programs to identify landslide-susceptible areas around the world. Other factors taken into account include topography, geology, roadways, fault zones, and deforestation. The system tracks the preceding seven days of rain and can be updated every thirty seconds. If rainfall is intense in a high-risk area, the system issues an alert for a high or moderate landslide risk.

But these systems and programs cannot predict exactly when a landslide will happen, how far it will go, or how it will behave. Scientists are particularly puzzled by how mudflows or rock avalanches move long distances at high velocities over relatively flat terrain. Add to the mix human-built structures such as roadways, homes, commercial buildings, and culverts, and the processes become even more complex. Another unknown is how climate change will influence the frequency and extent of landslides. More wildfires, heavier precipitation, and increased flooding could all cause an increase. Hopefully, as more is learned about landslides, more people will recognize the dangers they pose and be better warned in advance.

Rip Currents

HAULOVER PARK, FLORIDA, 2017. The refreshing onshore breeze, combined with warm air and sunny skies, made for a perfect beach day. It was May in South Florida. At Haulover Park, just north of Miami Beach, the water was clear and the waves relatively small—less than a meter high. A group of Orthodox rabbis visiting from New York City stopped by a beach tower to consult with lifeguards—red flags were flying. They were told of hazardous conditions and advised to swim near the lifeguard tower. But due to their religious beliefs, they did not want to swim near women. The rabbis chose to go elsewhere—an unpopulated and unguarded section of the beach.

Four men went into the water. Sadly, only two survived. They had entered the ocean at the exact spot where a narrow jet of seawater was flowing offshore—a rip current. There were no visible signs of the danger—no line of sediment-laced water, or foam, or change in color. Headlines across the world asserted that it had been a "strong rip current." But beach expert Stephen Leatherman disagrees, noting that strong rip currents happen in stormy conditions or on higher wave beaches. This one occurred on a day with clear water and relatively small waves. Rip currents don't have to be strong to be deadly.

Each year in the United States about a hundred people lose their lives in rip currents. Although no global database exists for rip current fatalities, a significant number of deaths have also been reported from the United Kingdom, India, Brazil, Australia, Costa Rica, and Turkey. Tens of thousands require rescue from rip currents annually. Over the last several decades, as technology has advanced, scientists have made significant progress in understanding how and where rip currents form, yet still there remain some watery wish-we-knews.

Rip currents are relatively narrow jets of seawater that flow offshore from a beach, generally extending seaward past the breaking waves. They may be weak or strong; tens of meters wide, or less; and their speed varies from a relatively slow 0.3 meters per second up to mega-rip velocities of faster than 2 meters per second—a current not even Olympic champion Michael Phelps could outswim. The formation of a rip current depends on a number of interacting factors, including wave height, the angle at which the waves strike the shore, seafloor bathymetry, and the presence of human-built structures such as jetties or seawalls.

Essentially, rip currents are the result of water piling up on a beach unevenly along its length. This may be due to differences in the height of breaking waves along shore or to the configuration of the shoreline. Once the water piles or is "set-up" on shore, it flows to low spots on the beach and then moves offshore in a rip, a fast, narrow flow. Rip currents are distinctly different from undertows or tidal jets (and please do not call them rip tides). While rip currents can be dangerous, they also play an important role in the transport and mixing of heat, nutrients, sediment, pollution, and marine species. During storms, rip currents can transport tremendous quantities of sediment offshore.

Rip currents are particularly hard to study due to their dynamic and ephemeral nature—and because humans and equipment don't do well in the surf zone, especially under high wave or storm conditions. Today, scientists use Acoustic Doppler Current Profilers and pressure sensors to study rip currents, as well as drifters outfitted with GPS and a harmless dye that turns bright green in seawater. These methods can also be combined with satellite imagery, space-based radar data, numerical modeling, and laboratory experiments. Unmanned aircraft or drones are also getting into the rip current game.

Scientists have discovered that there are different kinds of rip currents, distinguished by how they form, how long they last, and how they flow. Multiple rips can occur on a single beach. And rip currents can change quickly or strengthen in pulses. On San Diego beaches, rapidly changing rip currents have led to mass rescues. In 2011 twenty-six people were pulled safely from a strong rip current created by a tide change combined with the beginnings of a big incoming swell. In 2015 twenty-two swimmers were rescued from a quickly forming rip current on Mission Beach.

For years, scientists believed that all rip currents flowed offshore and dissipated seaward of the surf zone. More recent research has revealed that in some cases rip currents form circulating cells in which seawater flowing offshore turns parallel to shore and then heads back toward the beach. It is not clear why some rip currents form circulating cells while others are simple offshore jets. But this discovery has led to some debate on what to do if caught in a rip current. Everyone agrees on two things: do not panic and do not try to swim against the current directly back to shore—this is how most people drown. Swimmers should try to relax and float with the rip, call for help, and, if possible, swim parallel to shore to get out of the seaward flow. If it is a circulating flow, the current could help to bring a swimmer back to shore—but there's no guarantee or way of knowing.

The National Weather Service now issues rip current alerts based on numerical models that take into account winds, wave heights, tides, and long-period swells. At the beach, if rip currents are present, red flags may be posted at lifeguard stations. In some cases, rip currents can be seen and recognized from the beach. Swimmers should look for sediment-laden jets flowing offshore, lines of foam in a rip, or distinctly dark water. Places along the shore where there are fewer breaking waves may also indicate an offshore flow.

Rip currents are a common phenomenon. They tend to be stronger where the waves are higher or have a longer period (the time between crests). With typically higher-wave conditions, the US Pacific Coast is more prone to strong rip currents than beaches along the Gulf of Mexico, Atlantic Coast, or Great Lakes. Yet Florida has the highest number of rip current fatalities each year. In part, this is because people are less likely to enter the water in high-wave conditions, like on the Pacific Coast. Plus, when high waves combine with lots of loose sediment, rip currents can be more easily seen and avoided (figure 5.3). In Florida, rip currents can happen—as was the case at Haulover Park—on relatively calm days. In addition, many of the millions who visit Florida beaches each year have little to no knowledge about rip currents, their warning signs, or what to do if caught in one.

Given the challenges of working in the surf zone under ex-

Figure 5.3. Mega-rip currents extending more than 300 meters off a California beach. Courtesy Tom Cozad, Newport Beach, CA.

treme conditions, we know less about rip currents that occur under high-wave and storm conditions. Climate change is also at play here. Rising sea levels and increasing storm intensities may alter where and when rip currents form and how strong they are. Modifications to the coast due to development or shoreline engineering are another complicating factor, with the potential to alter the flow of water and sediment along shore as well as wave dynamics.

As with many of the Earth's other dangerous forces, with rip currents the key to saving lives is increased scientific understanding, timely warnings, and an educated public.

Sinkholes

For classic car lovers, the day will live in infamy—when the earth opened up and devoured eight rare Corvettes in mint condition. It was early morning on February 12, 2014, in Bowling Green, Kentucky. Luckily, the National Corvette Museum hadn't opened yet, so there were no visitors in the building's Sky Dome area when a 12-meter-wide crater emerged and the cars tumbled 9 meters down like toys.

The infamous car-swallowing sinkhole was later found to have been caused by the partial collapse of a connected large cave complex. But the Corvettes weren't the first pricey cars to fall victim to a sinkhole. In May 1981 a giant sinkhole in Winter Park, Florida, swallowed five Porsches from a repair shop—along with the deep end of an Olympic-size swimming pool, a three-bedroom home, two streets, and some 191,000 cubic meters of sediment.

Rarely do sinkholes result in fatalities. However, in February 2013 in Seffner, Florida, a sinkhole opened up one night under the home of a thirty-six-year-old man. His brother, also in the house, attempted a rescue but, tragically, was unable to pull the man from the collapsing, dirt-filled, 6-meter hole.

Sinkholes are essentially depressions or holes in the ground where there is no external surface drainage—water drains through the subsurface or underground. They can be less than a meter or hundreds of meters across and just as deep. In the United States, sinkholes most commonly occur in Florida, Texas, Kentucky, Alabama, Mississippi, Tennessee, and Pennsylvania.

Most sinkholes form in areas where the underlying rock can be dissolved by groundwater. As rainwater passes through the atmosphere and then flows through soil it absorbs carbon dioxide and becomes slightly acidic. Weakly acidic groundwater can dissolve certain types of rock, such as salt (in beds or domes), gypsum, and limestone. Areas underlain by limestone are especially susceptible to dissolution; the resulting landscapes, riddled with holes, sinkholes, caves, fissures, or even harboring underground rivers, are called karst. Some 48 kilometers from the sinkhole that swallowed the Corvettes in Kentucky is Mammoth Cave National Park—one of the world's largest caves in a karst landscape. Sinkholes also occur in the ocean, where they are known as blue holes.

Not all sinkholes collapse suddenly. Some begin as shallow depressions and grow gradually over time. Rock may be dissolved by water draining down from the surface, or from below by groundwater. Once enough space has been carved out or enough groundwater has drained away, the lack of support causes the overlying rock to collapse. Warning signs of an impending sinkhole collapse include slumping ground, leaning trees or fences, the development of small water-filled depressions or ponds, and structural cracks in buildings.

Scientists today use satellite imaging and radar to identify sinkholes and study karst landscapes. Underground cavities and caves are also being mapped through exploration and with ground-penetrating radar. Numerical and laboratory modeling have furthered our understanding of how sinkholes form and change with time. Still, it is not yet possible to forecast exactly what patch of ground will collapse and when. Heavy rains, hurricanes, and floods, but also droughts, can contribute to the formation of sinkholes. So can aging and failing underground pipes, changes to surface drainage, and withdrawals from aquifers or oil reservoirs. Thawing permafrost is now also causing more of the ground to give way. Evidence suggests that climate change will alter where, when, and how many sinkholes form, but what exactly the future holds remains, for now, another wish-we-knew.

Sharks

The movie *Jaws* played on our instinctual terror of being eaten—no one wants to be lunch. The suspense, blood, torn limbs, and creepy music added to the impact, and sharks became one of our worse fears. The influence of the movie also led Peter Benchley, on whose novel the film was based, to become a staunch advocate for sharks and conservation.

Sharks are not in fact the human-hunting, people-eating monsters *Jaws* portrayed; rather, they are efficient natural predators that play a critical role in the ocean ecosystem. Sharks keep prey species in check, winnow out the weak or diseased, and help to keep the marine food web in balance. But they can be dangerous. Like bears, tigers, and wolves, sharks are wild animals. When they are harassed, baited, or put on defense, or their territory is threatened, sharks can become aggressive. Sharks also taste by biting—their eyesight is limited, especially in murky water—so biting-by-mistake happens.

There is a lot we don't know but wish we knew about sharks. Where do sharks spend their days and nights? What are they doing? Where do they breed? And how do they feed under natural conditions? Marine biologist Sonny Gruber, who studied lemon sharks for more than four decades (and was part of a team that wrangled sharks for the James Bond movie *Thunderball*), noted that it's exceedingly rare to observe sharks feeding in nature —so their table manners remain somewhat mysterious.

Marine biologist Barbara Block and her colleagues have been tagging the ocean's top predators for decades. They use both satellite tags, which can transmit data when the sharks are at the surface or be programmed to pop off for retrieval, and acoustic tags, which must be located with a receiver. By tagging hundreds of great white sharks off California's coast, they have discovered a lot about these magnificent creatures, including an annual migration pattern and the location of a great white "hang out."

During the summer months, great white sharks cruise the central California coast and feast on the area's abundant seals and sea lions. Come winter they set out across the open ocean for Hawaii. Around the halfway point, many of the sharks make a several-week stopover in an open ocean area now known as the White Shark Café. Here, the sharks have been documented making short repeated "bounce" dives, from the surface down to some 450 meters. Is this behavior indicative of deep-sea hunting or part of a ritual leading to great white hookups—is it a feeding area or mating ground? Scientists and shark enthusiasts want to know.

In 2018 a month-long expedition headed to the White Shark Café aboard the research vessel *Falkor*. Ahead of the ship, the team led by Barbara Block tagged more than thirty sharks and deployed several saildrones (unmanned autonomous vehicles with a sail) outfitted with instruments to locate previously tagged sharks and study other species. Once the ship arrived on site, researchers located, tracked, and studied white sharks in the area, retrieved pop-off tags, and used a wide range of oceanographic sampling tools to collect data on nutrients, other marine species, currents, and water chemistry.

At the surface and from sensors aboard satellites, the White Shark Café region appears desertlike, with little in the way of marine life. Deep down, however, the researchers discovered a layer of nutrient-rich plant life that supports a surprisingly rich food web. Concentrated at this depth they found fish, squid, crustaceans, and jellyfish. The oceanographic data collected suggests that small-scale eddies mix the nutrient-rich waters and promote the high productivity in the region—leading to the surprising abundance and diversity of marine life. Clearly, the white sharks are attracted to the salty smorgasbord, but researchers still aren't sure if they come only to feed or if dining is combined with a little romance and it is also a hot spot for mating. The White Shark Café is now under consideration for protection as a United Nations World Heritage Site to prevent fishing in this, a most unusual concentration of one of the ocean's greatest predators.

Because the ocean is so vast and deep, and most of it is cold, dark, and under high pressure, there remain many unknowns in the sea: about its creatures, seafloor, chemistry, currents, and geology, as well as the impacts of climate change. An entire series of books could be written about all of the wish-we-knews in the ocean, sharks included.

Sharks are wild and powerful animals that should always be treated with respect and a healthy dose of caution. To reduce your chances of encountering a shark in feeding mode, sage advice is not to swim at dusk or dawn, if the water is murky, or when schools of baitfish are present. Don't swim where people are fishing or if there are many seals or sea lions in the water, and it is better to swim in a group. And remember, statistics indicate that you are more likely to be killed by lightning, balloons, falling from a ladder, or using a toaster or lawnmower than by a shark. In a world of dangerous forces and unknown science, sharks are scary, but way less worrisome than climate change, hurricanes, volcanoes, earthquakes, landslides, and rip currents.

Conclusion

Knowing Enough to Act

Living on planet Earth is a mixed bag. It comes with conditions that make life possible and provide for wondrous beauty. But our home orb also presents many dangers that can be life-threatening, costly, and destructive. Science is key to understanding these threats and knowing how to protect lives, property, and economic stability.

While unknowns remain, there is no question that we know more now than ever before about the Earth's dangerous forces and phenomena, enough to determine the risks involved and to make wiser decisions accordingly. We know Earth's climate is warming at an accelerated and unnatural pace and that climate change is causing glaciers and ice sheets to melt, sea level to rise, and weather-related impacts to be more extreme. We know why most volcanoes erupt, how to monitor their activity, and in many cases when and how to issue critical warnings and order evacuations. We know where earthquakes are most likely to occur, who is at greatest risk, and what types of buildings are most vulnerable to shaking and collapse. We know where tsunamis have happened before and will happen again. And we can track hurricanes as never before and reliably forecast landfall days in advance.

While mysteries remain, we know enough to direct investments and shape policies that can save lives.

Why then are so many of the world's citizens still vulnerable and at risk? The answer is complex and involves, at the very least, money, politics, development, lack of information or understanding, conflict, and, sadly, human nature.

In places where sufficient food, clean water, or health care are unavailable or limited, planning for a disaster that may not come for tens or hundreds of years is unlikely to be a priority, and may be impossible without outside assistance. Civil unrest or conflict can pose further obstacles. Economic considerations figure heavily as nations, states, communities, and individuals must decide what is acceptable risk? Is the cost of retrofitting a school in a tsunami zone too high when the next major tsunami may not occur for several generations, but could also hit tomorrow? What are we willing to pay to protect our children, ourselves, and our communities? If people have not experienced an extreme event, such as a major earthquake or hurricane, in their or their parents' lifetime, they may not understand or even recognize the threats they face. And even those who have experienced such events may nonetheless deny the risks, or simply feel helpless.

Humans by nature are creatures of the short term, especially today when gratification comes from nearly instantaneous likes and retweets. Planning and investing for the future and events that may not come to pass in our lifetime is not something we are historically good at.

Much has been said about the politicization of science. The bottom line is this: science should not be based on or guided by politics. It should be based on facts, evidence, observations—*data*. Debate and discussion are hallmarks of good science, but they must be based on credible data. Ideology, self-serving agendas, electoral concerns, and so on should not dictate policies and investments having to do with science or science-based issues. When political leaders and information providers promote policies or actions based simply on their beliefs or motives, they are putting thousands, if not millions, of people at risk, and our economy as well.

Scientists have long focused their efforts on research to better understand the world around us. Today, based on data and what we now know, scientists are increasingly speaking out about the urgent need for improved preparedness and investment in better warning systems, education, and mitigation, and for science-based policies. Earthquake experts like Robert Yeats, Ross Stein, and Lucy Jones are pleading with politicians and community leaders to invest in earthquake preparedness, retrofit buildings, and promote education for people, especially in high-risk areas. Volcanologists such as Robert Tilling, Michael Poland, and Christina Neal are working hard to help people prepare for volcanic events and to improve monitoring of volcanoes.

Oregon State University is planning to construct a new marine studies building on the Oregon coast—in a tsunami inundation zone. University leaders claim the building will be engineered to survive a 9.0 magnitude earthquake and the resultant tsunami. Even if the building survives—what about the people inside, who would still need to evacuate and be put at risk? Marine geologist Chris Goldfinger has spoken out against his own university's plans, emphasizing that they should exemplify tsunami preparedness, not build where others are being told not to. What if the university's construction encourages others to build in a danger zone?

Much like evolution and tobacco, climate change has become one of the most highly politicized, highly polarizing topics of our time. Scientists such as James Hansen and Michael Mann have spoken out against those who seek to dismiss the scientific evidence or argue, without credible data, against human-induced climate change. These well-respected researchers have been mercilessly attacked and harassed, even receiving death threats. This should not be how science is conducted or debated. Based on the ever-growing mass of data and very visible evidence, it is clear that the Earth's climate is changing at an accelerated pace and in a way that has not happened for hundreds of thousands of years, and just as clear that we are to blame. Have we reached a tipping point where it is too late to change the tide of ice melting, sea level rise, deadly heat waves, extreme flooding, and more severe storms? That remains a wish-we-knew, and an extraordinarily scary one. But based on the data, we know enough to act—and to act now—to reduce and mitigate the alarming reality that is climate change.

When it comes to hurricanes, many meteorologists are urging the public to be better prepared and to heed warnings, and those who control funding to continue to invest in weather and climate forecasting systems in order to improve how we communicate with and warn the public.

It is not the science unknowns these experts fear. It is the known—the well-documented dangers and the risks that we understand more clearly than ever but have not adequately addressed. They fear our lack of investment in studying and preparing for events that each year kill thousands of people and cost billions of dollars in damage. As the human population of our planet continues to rise, ever more people will be impacted by extreme and hazardous events. Billion-dollar disasters are becoming frequent—are trillion-dollar events on the horizon?

For now, we have only one planetary home—Earth. It is a dynamic, beautiful, and wondrous place. It is also fraught with powerful, unpredictable phenomena that can kill and destroy what we hold most dear. We must continue to invest in understanding our planet's dangerous forces, while also using what we have already learned to make the Earth a safer place now and for future generations.

Acknowledgments

My sincerest gratitude goes to all of those who have encouraged, supported, and inspired the writing of this book. That includes family, friends, colleagues, acquaintances, and even some strangers. Special thanks to Dave Jones for his endless support, humor, and advice; you are more appreciated than I can ever say or write. Steadfast and much appreciated encouragement was provided by my sister, Kathy Conrad, and by many good friends, including Linda Glover, Cathy Sherrill, and Peg Brandon, along with swimming pals Mary-Ellen and Rick Coles. Huge thanks to Christie Henry, executive editor extraordinaire, whose enthusiastic response to the idea for the book was a heavenly drink in an often difficult and disheartening publishing world. Though she moved on to new adventures during the writing of the book, her contribution cannot be overlooked and is very much appreciated. Thankfully, her replacement, Scott Gast, has been a pleasure to work with, and I am ever grateful for his keen advice and efforts. Much appreciation also goes to editor Joel Score for his thorough and exceptionally thoughtful work. I am especially grateful for his ability to catch the errors and inaccuracies that sometimes slip my notice. Thank you. Additional gratitude to all those at the

University of Chicago Press who work so hard to bring a book to completion. It definitely takes a team.

And finally, I want to express my great appreciation to all of the scientists who took the time to share their work and thoughts with me in person, on the phone, or via email. Your input to the book was invaluable, your work fascinating and admirable. A very big thank you goes to Bob Tilling, Bob Corell, Nancy Maynard, Don Swanson, Robert Yeats, Michael Poland, Michael Mann, Tim Dixon, Jacki Dixon, Ted Scambos, Jim White, Kevin Trenberth, Dennis Hansell, Rob Rogers, Frank Marks, Chris Langdon, Amy Clement, Bryan Norcross, Max Mayfield, Dave Jones, Chris Goldfinger, Ross Stein, Costas Synolakis, Eric Geist, Matthew Patrick, Christina Neal, Larry Mayer, John Clarke, Rick Thoman, Stephen Leatherman, Dave Petley, Beth Grassi, Brian Atwater, Patrick Dobson, Richard Styron, Peter Frenzen, Tom Casadevall, Margaret Leinen, Jeff Masters, Andreas Muenchow, Mark Serreze, Peter Lippman, Jacob Lowenstern, Yan Lavallee, David Jackson, Kerry Sieh, Richard Alley, Dennis Hubbard, Scott Alan Ashford, Josh Willis, Marshall Shephard, Les Kaufman, Frank Marks, Neal Dorst, Larry Peterson, John Bruno, Mitch Goldberg, Bill Lapenta, Don Chambers, Al Hine, Gary Mitchum, Alain Bonneville, Dave Schneider, and Kris Holderied. To anyone I missed, my apologies and thank you!

Readings and References

Climate Change

Abatzoglou, J. T., and A. P. Williams. "Impact of Anthropogenic Climate Change on Wildfires across Western US Forests." *Proceedings of the National Academy of Sciences* 113, no. 42 (2016): 11770–75. https://doi.org/10.1073/pnas.1607171113

Adhikari, S., E. R. Ivins, and E. Larour. "Mass Transport Waves Amplified by Intense Greenland Melt and Detected in Solid Earth Deformation." *Geophysical Research Letters* 44 (2017): 4965–75. https://doi.org/10.1002/2017GL073478

Alley, Karen F., Ted A. Scambos, Matthew R. Siegfried, and Helen Amanda Fricker. "Impacts of Warm Water on Antarctic Ice Shelf Stability through Basal Channel Formation." *Nature Geoscience* 9 (2016): 290–93. https://doi.org/10.1038/ngeo2675

Alley, R. B., J. Marotzke, W. D. Nordhaus, et al. "Abrupt Climate Change." *Science* 299, no. 5615 (2003): 2005–10. https://doi.org/10.1126/science.1081056

An, L., E. Rignot, S. Elieff, et al. "Bed Elevation of Jakobshavn Isbræ, West Greenland, from High-Resolution Airborne Gravity and Other Data." *Geophysical Research Letters* 44 (2017): 3728–36. https://doi.org/10.1002/2017GL073245

Antarctic Glaciers. "Explaining the Science of Antarctic Glaciology." http://www.antarcticglaciers.org

Arctic Institute. "Alaskan Villages Imperiled by Global Warming Need Resources to Relocate." July 2015. http://www.thearcticinstitute.org/alaskan-villages-imperiled-global-warming/

Arctic Monitoring and Assessment Programme. *Snow, Water, Ice and Permafrost: Summary for Policy-makers*. Oslo: AMAP, 2017.

Arctic Monitoring and Assessment Programme. "Snow, Water, Ice and Permafrost in the Arctic (SWIPA), 2017." https://www.amap.no/documents/doc/Snow-Water-Ice-and-Permafrost-in-the-Arctic-SWIPA-2017/1610

Barletta, Valentina R., Michael Bevis, Benjamin E. Smith, et al. "Observed Rapid Bedrock Uplift in Amundsen Sea Embayment Promotes Ice-Sheet Stability." *Science* 360, no. 6395 (2018): 1335–39. https://doi.org/10.1126/science.aao1447

Barton, A. D, A. J. Irwin, Z. V. Finkel, and C. A. Stocka. "Anthropogenic Climate Change Drives Shift and Shuffle in North Atlantic Phytoplankton Communities." *Proceedings of the National Academy of Sciences* 113, no. 11 (2016): 2964–69. https://doi.org/10.1073/pnas.1519080113

British Antarctic Survey. "Discovering Antarctica." http://discovering antarctica.org.uk

British Broadcasting Company. "In Siberia There Is a Huge Crater and It Is Getting Bigger." http://www.bbc.com/earth/story/20170223-in-siberia-there-is-a-huge-crater-and-it-is-getting-bigger

Broecker, W. S. "Does the Trigger for Abrupt Climate Change Reside in the Ocean or in the Atmosphere?" *Science* 300, no. 5625 (2003): 1519–22. https://doi.org/10.1126/science.1083797

Cable News Network. "Tragedy of a Village Built on Ice." http://www.cnn.com/2017/03/29/us/sutter-shishmaref-esau-tragedy/index.html

Chadburn, S. E., E. J. Burke, P. M. Cox, et al. "An Observation-Based Constraint on Permafrost Loss as a Function of Global Warming." *Nature Climate Change* 7 (2017): 340–45. https://doi.org/10.1038/NCLIMATE3262

Chen, I-Ching, J. K. Hill, R. Ohlemüller, et al. "Rapid Range Shifts of Species Associated with High Levels of Climate Warming." *Science* 333, no. 6045 (2011): 1024–26. https://doi.org/10.1126/science.1206432

Cheng, L., K. E. Trenberth, J. Fasullo, J. Abraham, et al. "Taking the Pulse of the Planet." *Eos* 98 (2017). https://doi.org/10.1029/2017EO081839

Cheng, L., K. E. Trenberth, J. Fasullo, T. Boyer, et al. "Improved Estimates of Ocean Heat Content from 1960 to 2015." *Science Advances* 3 e1601545 (2017). https://doi.org/10.1126/sciadv.1601545

Clement, A. C., and L. C. Peterson. "Mechanisms of Abrupt Climate Change of the Last Glacial Period." *Reviews of Geophysics* 46, RG4002 (2008). https://doi.org/10.1029/2006RG000204

Climate Central. "Atlantic Circulation Weaker Than in Last Thousand Years." http://www.climatecentral.org/news/climate-change -jamming-critical-heat-conveyor-18810

Climate Central. "Climate Change's Evolving Role in Extreme Weather." http://www.climatecentral.org/news/climate-change-evolving-role -in-extreme-weather-18501

Climate Central. "East Antarctica Is Melting from Above and Below." http://www.climatecentral.org/news/east-antarctica-melting -climate-change-20986

Climate Central. "Larsen C Rift Is Racing to Its Conclusion." http://www .climatecentral.org/news/larsen-c-rift-speed-up-21577

Climate Communication. "Heat Waves and Climate Change." https:// www.climatecommunication.org/new/features/heat-waves-and -climate-change/overview/

Costanza, R., R. Groot, P. Sutton, et al. "Changes in the Global Value of Ecosystem Services." *Global Environmental Change* 26 (2014): 152– 58. https://doi.org/10.1016/j.gloenvcha.2014.04.002

DeConto, R. M., and D. Pollard. "Contribution of Antarctica to Past and Future Sea-Level Rise." *Nature* 531 (2016): 591–97. https://doi.org /10.1038/nature17145

Dixon, Timothy H. *Curbing Catastrophe: Natural Hazards and Risk Reduction in the Modern World.* Cambridge: Cambridge University Press, 2017.

Eakin, D. M., G. Liu, A. M. Gomez, et al. "Global Coral Bleaching 2014– 2017." *Reef Encounter* 31 (2016): 20–25.

Favier, L., G. Durand, S. L. Cornford, et al. "Retreat of Pine Island Glacier Controlled by Marine Ice-Sheet Instability." *Nature Climate Change* 4 (2014): 117–21. https://doi.org/10.1038/nclimate2094

Fenty, I., J. K. Willis, A. Khazendar, et al. "Oceans Melting Greenland: Early Results from NASA's Ocean-Ice Mission in Greenland." *Oceanography* 29, no. 4 (2016): 72–83. https://doi.org/10.5670/oceanog .2016.100

Frajka-Williams, E., J. L. Bamber, and K. Våge. "Greenland Melt and the Atlantic Meridional Overturning Circulation." *Oceanography* 29, no. 4 (2016): 22–33. https://doi.org/10.5670/oceanog.2016.96

Froese, Duane G., John A. Westgate, Alberto V. Reyes, et al. "Ancient Permafrost and a Future, Warmer Arctic." *Science* 321, no. 5896 (2008): 1648. http://doi.org/10.1126/science.1157525

Graeter, K. A., E. C. Osterberg, D. G. Ferris, et al. "Ice Core Records of

West Greenland Melt and Climate Forcing." *Geophysical Research Letters* 45 (2018). https://doi.org/10.1002/2017GL076641

Green, J. K., S. I. Seneviratne, A. M. Berg, et al. "Large Influence of Soil Moisture on Long-Term Terrestrial Carbon Uptake." *Nature* 565 (2019): 476–79.

Grist. "Ice Apocalypse." https://grist.org/article/antarctica-doomsday -glaciers-could-flood-coastal-cities/

Harvell, C. D., D. Montecina-Latorre, J. M. Caldwell, et al. "Disease Epidemic and Marine Heat Wave Are Associated with the Continental-Scale Collapse of a Pivotal Predator." *Science Advances* 5, no. 1 (2019): eaau7042. https://doi.org/10.1126/sciadv.aau7042

Hawken, Paul, ed. *Drawdown: The Most Comprehensive Plan Ever Proposed to Reverse Global Warming.* New York: Penguin, 2017.

Henry, L. G., J. F. McManus, W. B. Curry, et al. "North Atlantic Ocean Circulation and Abrupt Climate Change during the Last Glaciation." *Science* 353, no. 6298 (2016): 470–74. https://doi.org10.1126 /science.aaf5529

Hofer, Stefan, Andrew J. Tedstone, Xavier Fettweis, and Jonathan L. Bamber. "Decreasing Cloud Cover Drives the Recent Mass Loss on the Greenland Ice Sheet." *Science Advances* 3, no. 6 (2017): e1700584. https://doi.org/10.1126/sciadv.1700584

Hughes, T. P., J. T. Kerry, M. Álvarez-Noriega, et al. "Global Warming and Recurrent Mass Bleaching of Corals." *Nature* 543 (2017): 373–77. https://doi.org/10.1038/nature21707

Hughes, T. P., J. T. Kerry, A. H. Baird, et al. "Global warming impairs stock-recruitment dynamics of corals." *Nature* 568 (2019): 387–390. http://dx.doi.org/10.1038/s41586-019-1081-y

Inside Climate News. "Massive Permafrost Thaw Documented in Canada, Portends Huge Carbon Release." https://insideclimatenews.org /news/27022017/global-warming-permafrost-study-melt-canada -siberia

Inside Climate News. "Sea Level Rise Will Rapidly Worsen in Coming Decades, NOAA Warns." https://insideclimatenews.org/news /07032018/sea-level-rise-data-global-warming-noaa-coastal-cities -united-states-climate-change

Intergovernmental Panel on Climate Change. "Climate Change 2013: The Physical Science Basis." http://www.ipcc.ch/report/ar5/wg1/

Intergovernmental Science-Policy Platform on Biodiversity and Ecosystem Services. "Nature's Dangerous Decline 'Unprecedented' Species Extinction Rates 'Accelerating.'" 2019. https://www.ipbes.net/news /Media-Release-Global-Assessment

International Research Institute for Climate and Society, Columbia Uni-

versity. "Extreme Tornado Outbreaks Have Become More Common."
http://iri.columbia.edu/news/tornado-outbreaks/

Joughin, I., B. E. Smith, D. E. Shean, and D. Floricioiu. "Brief Communication: Further Summer Speedup of Jakobshavn Isbræ." *Cryosphere* 8 (2014): 209–14, https://doi.org/10.5194/tc-8-209-2014

Kerr, Richard A. "Arctic Armageddon Needs More Science, Less Hype." *Science* 329, no. 5992 (2010): 620–21. https://doi.org/10.1126 /science.329.5992.620

Kingslake, Jonathan, Jeremy C. Ely, Indrani Das, and Robin E. Bell. "Widespread Movement of Meltwater onto and across Antarctic Ice Shelves." *Nature* 544 (2017): 349. https://doi.org/doi:10.1038 /nature22049

Lacis, Andrew A., Gavin A. Schmidt, David Rind, and Reto A. Ruedy. "Atmospheric CO_2: Principal Control Knob Governing Earth's Temperature." *Science* 330, no. 6002 (2010): 356–59. https://doi.org/10.1126 /science.1190653

Larsen, C. F., E. Burgess, A. A. Arendt, et al. "Surface Melt Dominates Alaska Glacier Mass Balance." *Geophysical Research Letters* 42 (2015): 5902–8. https://doi.org/10.1002/2015GL064349

Lenaerts, J. T. M., S. Lhermitte, R. Drews, et al. "Meltwater Produced by Wind-Albedo Interaction Stored in an East Antarctic Ice Shelf." *Nature Climate Change* 7 (2017): 58–62. https://doi.org/10.1038 /nclimate3180

Li, X., E. Rignot, M. Morlighem, et al. "Grounding Line Retreat of Totten Glacier, East Antarctica, 1996 to 2013." *Geophysical Research Letters* 42 (2015): 8049–56. https://doi.org/10.1002/2015GL065701

Live Science. "Antarctica: Facts About the Coldest Continent." http:// www.livescience.com/21677-antarctica-facts.html

Los Angeles Times. "California's Camp Fire Was the Costliest Global Disaster Last Year, Insurance Report Shows." https://www.latimes.com /local/lanow/la-me-ln-camp-fire-insured-losses-20190111-story .html

MacGregor, J. A., M. A. Fahnestock, G. A. Catania, et al. "Radiostratigraphy and Age Structure of the Greenland Ice Sheet." *Journal of Geophysical Research Earth Surface* 120 (2015): 212–24. https://doi .org/10.1002/2014JF003215

Mann, M. E. *The Hockey Stick and the Climate Wars*. New York: Columbia University Press, 2012.

Mann, Michael E., Raymond S. Bailey, and Malcolm K. Hughes, "Northern Hemisphere Temperatures during the Past Millennium: Inferences, Uncertainties, and Limitations," *Geophysical Research Letters* 26, no. 6 (1999): 759–62.

Mann, Michael E., and Lee R. Kump. *Dire Predictions: Understanding Climate Change*. 2nd ed. New York: DK Publishing, 2015.

Mann, Michael E., Stefan Rahmstorf, Kai Kornhuber, et al. "Influence of Anthropogenic Climate Change on Planetary Wave Resonance and Extreme Weather Events." *Scientific Reports* 7 (2017). https://doi.org/10.1038/srep45242

Massom, Robert A., Theodore A. Scambos, Luke G. Bennetts, et al. "Antarctic Ice Shelf Disintegration Triggered by Sea Ice Loss and Ocean Swell." *Nature* (2018). https://doi.org/10.1038/s41586-018-0212-1

Mauritsen, T., and R. Pincus. "Committed Warming Inferred from Observations." *Nature Climate Change* 7 (2017): 652–55. https://doi.org/10.1038/nclimate3357

Mazdiyasni1, O., A. AghaKouchak1, S. J. Davis, et al. "Increasing Probability of Mortality during Indian Heat Waves." *Science Advances* 3, no. 6 (2017): e1700066. https:/doi.org/10.1126/sciadv.1700066

McManus, J. F., R. Francois, J. M. Gherardi, et al. "Collapse and Rapid Resumption of Atlantic Meridional Circulation Linked to Deglacial Climate Changes." *Nature* 428 (2004): 834–37. https//doi.org/10.1038/nature02494

Mikkelsen, Naja, and Torsten Ingerslev. "Ilulissat Icefjord: Nomination Document for the World Heritage Site." Copenhagen: Geological Survey of Denmark and Greenland, 2018.

Mora, C., B. Dousset, I. R. Caldwell, et al. "Global Risk of Deadly Heat." *Nature Climate Change* 7 (2017): 501–6. https://doi.org/10.1038/nclimate3322

Morlighem, M., E. Rignot, and J. K. Willis. "Improving Bed Topography Mapping of Greenland Glaciers Using NASA's Oceans Melting Greenland (OMG) Data." *Oceanography* 29, no. 4 (2016): 62–71. https://doi.org/10.5670/oceanog.2016.99

Motyka, R. J., M. Truffer, M. Fahnestock, et al. "Submarine Melting of the 1985 Jakobshavn Isbræ Floating Tongue and the Triggering of the Current Retreat." *Journal Geophysical Research* 116 (2011): F01007. https://doi.org/10.1029/2009JF001632

Münchow, A., L. Padman, P. Washam, and K. W. Nicholls. "The Ice Shelf of Petermann Gletscher, North Greenland, and Its Connection to the Arctic and Atlantic Oceans." *Oceanography* 29, no. 4 (2016): 84–95. https://doi.org/10.5670/oceanog.2016.101

National Academy of Sciences, Engineering, Medicine. "Attribution of Extreme Weather Events in the Context of Climate Change." https://www.nap.edu/resource/21852/Attribution-Extreme-Weather-Brief-Final.pdf

National Aeronautics and Space Administration. "Breakup of the Larsen Ice Shelf, Antarctica." https://earthobservatory.nasa.gov/images /2288

National Aeronautics and Space Administration. "Fastest Glacier in Greenland Doubles Speed." https://www.nasa.gov/vision/earth /lookingatearth/jakobshavn.html

National Aeronautics and Space Administration. "Huge Cavity in Antarctic Glacier Signals Rapid Decay." https://www.jpl.nasa.gov/news /news.php?feature=7322

National Aeronautics and Space Administration. "NASA Data Peers into Greenland's Ice Sheet." https://www.nasa.gov/content/goddard /nasa-data-peers-into-greenlands-ice-sheet

National Aeronautics and Space Administration. "NASA Discovers a New Mode of Ice Loss in Greenland." https://climate.nasa.gov/news /2591/nasa-discovers-a-new-mode-of-ice-loss-in-greenland/

National Aeronautics and Space Administration. "NASA's OMG Mission Maps Greenland's Coastline." https://www.nasa.gov/jpl/nasas-omg -mission-maps-greenlands-coastline

National Aeronautics and Space Administration. "Q&A with NASA's Joey Comiso: What Is Happening with Antarctic Sea Ice?" https:// www.nasa.gov/content/goddard/qa-what-is-happening-with -antarctic-sea-ice/

National Oceanic and Atmospheric Administration. "Billion-Dollar Weather and Climate Disasters." https://www.ncdc.noaa.gov/billions /overview

National Oceanic and Atmospheric Administration. "Climate Change: Global Temperature." https://www.climate.gov/news-features /understanding-climate/climate-change-global-temperature

National Oceanic and Atmospheric Administration. "Coral Bleaching during & since the 2014–2017 Global Coral Bleaching Event Status and an Appeal for Observations." https://coralreefwatch.noaa.gov /satellite/analyses_guidance/global_coral_bleaching_2014-17 _status.php

National Oceanic and Atmospheric Administration. "What Are Isotopes of Greenhouse Gases?" https://www.esrl.noaa.gov/gmd/infodata /faq_cat-3.html#5

National Public Radio. "An Ice Shelf Is Cracking in Antarctica, but Not for the Reason You Think." http://www.npr.org/sections/thetwo -way/2017/01/16/509565462/an-ice-shelf-is-cracking-in-antarctica -but-not-for-the-reason-you-think

National Snow and Ice Data Center. "Climate and Frozen Ground." https://nsidc.org/cryosphere/frozenground/climate.html

National Snow and Ice Data Center. "Methane and Frozen Ground." https://nsidc.org/cryosphere/frozenground/methane.html

National Snow and Ice Data Center. "State of the Cryosphere: Permafrost and Frozen Ground." https://nsidc.org/cryosphere/sotc/permafrost.html

Natural Environment Research Council. "Changing Arctic Ocean: Implications for Marine Biology and Biochemistry." http://www.nerc.ac.uk/research/funded/programmes/arcticocean/#xcollapse5

Nerem, R. S., B. D. Beckley, J. T. Fasullo, et al. "Climate-Change–Driven Accelerated Sea-Level Rise." *Proceedings of the National Academy of Sciences* 115, no. 9 (2018): 2022–25. https://doi.org/10.1073/pnas.1717312115

New York Times. "Europe Was Colder Than the North Pole This Week. How Could That Be?" https://www.nytimes.com/2018/03/01/climate/polar-vortex-europe-cold.html

Nicolas, Julien P., Andrew M. Vogelmann, Ryan C. Scott, et al. "January 2016 Extensive Summer Melt in West Antarctica Favoured by Strong El Niño." *Nature Communications* 8, no 15299 (2017). https://doi.org/10.1038/ncomms15799

NOAA National Centers for Environmental Information (NCEI). "U.S. Billion-Dollar Weather and Climate Disasters (2017)." https://www.ncdc.noaa.gov/billions/

Pachauri, R. K., and L. A. Meyer, eds. *Climate Change 2014: Synthesis Report*. Contribution of Working Groups I, II, and III to the Fifth Assessment Report of the Intergovernmental Panel on Climate Change. Geneva: IPCC, 2014.

Penn State. "Extreme Weather Events Linked to Climate Change Impact on the Jet Stream." https://news.psu.edu/story/458049/2017/03/27/research/extreme-weather-events-linked-climate-change-impact-jet-stream

Petermann's Glacial History. "Deciphering Past Climate at the Ice-Ocean Boundary." https://petermannsglacialhistory.wordpress.com

Pinsky, M. L., B. Worm, M. J. Fogarty, et al. "Marine Taxa Track Local Climate Velocities." *Science* 341, no. 6151 (2013): 1239–42. https://doi.org/10.1126/science.1239352

Project Midas. "Larsen C Iceberg Accelerates ahead of Calving." http://www.projectmidas.org/blog/berg-acceleration/

Qui, Jane. "Why Slow Glaciers Can Sometimes Surge as Fast as a Speeding Train—Wiping Out People in Their Path." *Science* (November 2017). https://doi.org/10.1126/science.aar5965

Raftery, A. E., A. Zimmer, D. M. W. Frierson, et al. "Less Than 2 °C

Warming by 2100 Unlikely." *Nature Climate Change* 7 (2017): 637–41. https://doi.org/10.1038/nclimate3352

Rahmstorf, Stefan, Jason E. Box, Georg Feulner, et al. "Exceptional Twentieth-Century Slowdown in Atlantic Ocean Overturning Circulation." *Nature Climate Change* 5 (2015): 475–80. https:/doi.org/10.1038/nclimate2554

Renssen, Hans, Aurélien Mairesse, Hugues Goosse, et al. "Multiple Causes of the Younger Dryas Cold Period." *Nature Geoscience* 8 (2015): 946–49. https://doi.org/10.1038/ngeo2557

Rignot, E., J. Mouginot, and B. Scheuch. "Ice Flow of the Antarctic Ice Sheet." *Science* 333, no. 6048 (2011): 1427–30. https://doi.org/10.1126/science.1208336

Rignot, E., J. Mouginot, M. Morlighem, et al. "Widespread, Rapid Grounding Line Retreat of Pine Island, Thwaites, Smith, and Kohler Glaciers, West Antarctica, from 1992 to 2011." *Geophysical Research Letters* 41 (2014): 3502–9. https://doi.org/10.1002/2014GL060140

Rintoul, Stephen Rich, Alessandro Silvano, Beatriz Pena-Molino, et al. "Ocean Heat Drives Rapid Basal Melt of the Totten Ice Shelf." *Science Advances* 2, no. 12 (December 2016): e1601610. https://doi.org/10.1126/sciadv.1601610

Rolling Stone. "Welcome to the Age of Climate Migration." https://www.rollingstone.com/politics/news/welcome-to-the-age-of-climate-migration-w516974

Romm, Joseph. *Climate Change: What Everyone Needs to Know.* New York: Oxford University Press, 2015.

Scambos, T. "Detailed Ice Loss Pattern in the Northern Antarctic Peninsula: Widespread Decline Driven by Ice Front Retreats." *Cryosphere* 9 (2014): 2135–45.

Scambos, T. A., G. G. Campbell, A. Pope, et al. "Ultra-Low Surface Temperatures in East Antarctica from Satellite Thermal Infrared Mapping: The Coldest Places on Earth." *Geophysical Research Letters* (June 2018). https://doi.org/10.1029/2018GL078133

Schmidt, M. W., and J. E. Hertzberg. "Abrupt Climate Change during the Last Ice Age." *Nature Education Knowledge* 3, no. 10 (2011): 11.

Schmidtko, S., L. Stramma, and M. Visbeck. "Decline in Global Oceanic Oxygen Content during the Past Five Decades." *Nature* 542 (2017): 335–39. https://doi.org/10.1038/nature21399

Schmidtko, Sunke, Karen J. Heywood, Andrew F. Thompson, and Shigeru Aoki. "Multidecadal Warming of Antarctic Waters." *Science* 346, no. 6214 (2014): 1227–31. https://doi.org/10.1126/science.1256117

Schroeder, D., A. Hilger, J. Paden, et al. "Ocean Access beneath the Southwest Tributary of Pine Island Glacier, West Antarctica." *Annals of Glaciology* (2017): 1–6. https://doi.org/10.1017/aog.2017.45

Science Daily. "Ocean Circulation Implicated in Past Abrupt Climate Changes." https://www.sciencedaily.com/releases/2016/06/160630145000.htm

Scientific American. "Alaska's Glaciers Are Retreating." https://www.scientificamerican.com/article/alaska-s-glaciers-are-retreating/

Scientific American. "Fresh Mammoth Carcass from Siberia Holds Many Secrets." https://www.scientificamerican.com/article/fresh-mammoth-carcass-from-siberia-holds-many-secrets/

Scientific Committee on Antarctic Research. https://scar.org

Shepherd, Andrew, Erik Ivins, Eric Rignot, et al. "Mass Balance of the Antarctic Ice Sheet from 1992 to 2017." *Nature* 558 (2081): 219–22.

Siegert, M. J., N. Ross, and A. M. Le Brocq. "Recent Advances in Understanding Antarctic Subglacial Lakes and Hydrology." *Philosophical Transactions A: Mathematical, Physical, and Engineering Sciences* 374, no. 2059 (2016): 20140306. https://doi.org/10.1098/rsta.2014.0306

Silvano, A., S. R. Rintoul, and L. Herraiz-Borreguero. "Ocean–Ice Shelf Interaction in East Antarctica." *Oceanography* 29, no.4 (2016): 130–43. https://doi.org/10.5670/oceanog.2016.105

Silvano, W. A., S. R. Rintoul, B. Peña-Molino, et al. "Freshening by Glacial Meltwater Enhances Melting of Ice Shelves and Reduces Formation of Antarctic Bottom." *Science Advances* 4, no. 4 (2018): eaap9467. https://doi.org/10.1126/sciadv.aap9467

Stanley, S. "Recent Studies Crack Open New Views of Glacial Crevasses." *Eos* 97 (2016). https://doi.org/10.1029/2016EO047897

Steffensen, Jørgen Peder, Katrine K. Andersen, Matthias Bigler, et. al. "High-Resolution Greenland Ice Core Data Show Abrupt Climate Change Happens in Few Years." *Science* 321, no. 5889 (2008): 680–84. https://doi.org/10.1126/science.1157707

Straneo, F., G. S. Hamilton, L. A. Stearns, and D. A. Sutherland. "Connecting the Greenland Ice Sheet and the Ocean: A Case Study of Helheim Glacier and Sermilik Fjord." *Oceanography* 29, no. 4(2016): 34–45. https://doi.org/10.5670/oceanog.2016.97

Sweet, W. V., R. E. Kopp, C. P. Weaver, et al. "Global and Regional Sea Level Rise Scenarios for the United States." NOAA Technical Report NOS CO-OPS 083. NOAA/NOS Center for Operational Oceanographic Products and Services (2017).

Tedesco, M., S. Doherty, X. Fettweis, et al. "The Darkening of the Greenland Ice Sheet: Trends, Drivers, and Projections (1981–2100)." *Cryosphere* 10 (2016): 477–96. https://doi.org/10.5194/tc-10-477-2016

Thornally, D. J. R., D. W. Oppo, P. O. Ortega, et al. "Anomalously Weak Labrador Sea Convection and Atlantic Overturning during the Past 150 Years." *Nature* 556 (2018): 227–30. https://doi.org/10.1038/s41586-018-0007-4

Tollefson, J. "Satellite Snafu Masked True Sea-Level Rise for Decades." *Nature* 547, no. 7663 (2017): 265–66. https://doi.org/10.1038/nature.2017.22312

Trenberth, K. E., J. T. Fasullo, and T. G. Shepherd. "Attribution of Climate Extreme Events." *Nature Climate Change* 5 (2015): 725–730. https:/doi.org/10.1038/nclimate2657

Tripati, Aradhna K., Christopher D. Roberts, and Robert A. Eagle. "Coupling of CO_2 and Ice Sheet Stability over Major Climate Transitions of the Last 20 Million Years." *Science* 326, no. 5958 (2009): 1394–97. https://doi.org/10.1126/science.1178296

United States Government Accountability Office. "Alaskan Villages Under Threat." http://www.gao.gov/new.items/d09551.pdf

Washington Post. "Greenland Is Melting Faster Than at Any Time in the Last 450 Years at Least." https://www.washingtonpost.com/news/energy-environment/wp/2018/03/28/greenland-is-melting-faster-than-at-any-time-in-the-last-450-years-at-least/

Washington Post. "Scientists Confirm Their Fears about West Antarctica—That It's Inherently Unstable." https://www.washingtonpost.com/news/energy-environment/wp/2015/11/02/scientists-confirm-their-fears-about-west-antarctica-that-its-inherently-unstable/

Washington Post. "Scientists Mapping Greenland Have Produced Some Surprising and Worrying Results." https://www.washingtonpost.com/news/energy-environment/wp/2017/10/04/scientists-mapping-greenland-have-produced-some-surprising-and-worrying-results/

Washington Post. "Scientists Say Greenland's Vast Melt Hasn't Slowed Down the Atlantic Ocean's Circulation—Yet." https://www.washingtonpost.com/news/energy-environment/wp/2016/06/20/a-huge-science-debate-is-brewing-over-whether-weve-messed-up-the-atlantic-oceans-circulation/

Wdowinski, S., R. Bray, B. P. Kirtman, and Z. Wu. "Increasing Flooding Hazard in Coastal Communities Due to Rising Sea Level: Case Study of Miami Beach, Florida." *Ocean & Coastal Management* 126 (2016): 1–8.

Weather Underground. "The Science of Abrupt Climate Change: Should

We Be Worried?" https://www.wunderground.com/resources
/climate/abruptclimate.asp

Wendish, M., M. Brückner, J. P. Burrows, et al. "Understanding the
Causes and Effects of Rapid Warming in the Arctic." *Earth & Space Science News* 98, no. 8 (2017): 22–26.

Wuebbles, D. J., D. W. Fahey, K. A. Hibbard, et al., eds. "USGCRP, 2017:
Climate Science Special Report: Fourth National Climate Assessment." Vol. 1. US Global Change Research Program, Washington, DC,
2017. https://doi.org/doi: 10.7930/J0J964J6

Yale Environment 360. "How Climate Change Could Jam the World's
Ocean Circulation." http://e360.yale.edu/features/will_climate
_change_jam_the_global_ocean_conveyor_belt

Volcanoes

Babb, J. L., J. P. Kauahikaua, and R. I. Tilling. *The Story of Hawaiian Volcano Observatory: A Remarkable First 100 Years of Tracking Eruptions and Earthquakes*. US Geological Survey, General Information Product
135, 2011.

Cable News Network. "BBC Crew Makes Dramatic Escape as Mount Etna
Volcano Erupts." http://www.cnn.com/2017/03/17/europe/bbc
-crew-volcano-mount-etna-eruption/index.html

Chadwick, W. W., Jr., B. P. Paduan, D. A. Clague, et al. "Voluminous Eruption from a Zoned Magma Body after an Increase in Supply Rate at
Axial Seamount." *Geophysical Research Letters* 43 (2016): 12,063–70.
https:/doi.org/10.1002/2016GL071327

Colón, D. P., I. N. Bindeman, and T. V. Gerya. "Thermomechanical Modeling of the Formation of a Multilevel, Crustal-Scale Magmatic System
by the Yellowstone Plume." *Geophysical Research Letters* 45, no. 9
(2018): 3873–79. https://doi.org/10.1029/2018GL077090

Crandell, Dwight R., and Donal R. Mullineaux. "Potential Hazards from
Future Eruptions of Mount St. Helens Volcano, Washington." *Geological Survey Bulletin* 1383-C (1978).

Crandell, Dwight R., Donal R. Mullineaux, and Meyer Rubin. "Mount St.
Helens Volcano: Recent and Future Behavior." *Science* 187 (1975):
438–40.

Decker, Robert, and Barbara Decker. *Volcanoes*. New York: W. H. Freeman, 1979.

Earth magazine. "Where Fire Freezes: All Eyes, Ears and Instruments on
Iceland's Volatile Volcanoes." https://www.earthmagazine.org/article
/where-fire-freezes-all-eyes-ears-and-instruments-icelands-volatile
-volcanoes

Earthscope. "The Multi-Chambered Heart of Mount St. Helens." http://
www.earthscope.org/articles/The_Multi_Chambered_Heart_of
_Mount_St_Helens

Eurek Alert. "Ring around Bathtub at Giant Volcano Field Shows Movement of Subterranean Magma." https://www.eurekalert.org/pub
_releases/2018-06/uow-a062518.php

Forbes. "How Many Jelly Beans Would Fit inside Yellowstone's Magma
Chamber." https://www.forbes.com/sites/robinandrews/2017/10
/22/how-many-jelly-beans-would-fit-inside-yellowstones-magma
-chamber/

FUTUREVOLC. "A New Research Article in Science: Gradual Caldera Collapse at Bárdarbunga Volcano, Iceland, Regulated by Lateral Magma
Outflow." http://futurevolc.hi.is/new-research-article-science
-gradual-caldera-collapse-bárdarbunga-volcano-iceland-regulated
-lateral

FUTUREVOLC. "Exploiting the Outcome of FUTUREVOLC." http://
futurevolc.hi.is/sites/futurevolc.hi.is/files/Pdf/vedurstofan
_futurevolc_baeklingur.pdf

Geiger, H., T. Mattsson, F. M. Deegan, et al. 2016. "Magma Plumbing
for the 2014–2015 Holuhraun Eruption, Iceland." *Geochemistry,
Geophysics, Geosystems* 17 (2016): 2953–68. https//doi.org/10.1002
/2016GC006317

Geological Society of London. "Super-Eruptions: Global Effects and
Future Threats." http://pages.mtu.edu/~raman/VBigIdeas/Super
eruptions_files/Super-eruptionsGeolSocLon.pdf

Geology.com. "Loihi Seamount: The Next Volcanic Island in the Hawaiian Chain." http://geology.com/usgs/loihi-seamount/

Gizmodo. "Scientists Just Discovered Something Extraordinary about
Iceland's Huge Volcano." https://gizmodo.com/scientists-just
-discovered-something-extraordinary-abou-1783674109

Glicken, H. *Rockslide-Debris Avalanche of May 18, 1980, Mount St. Helens
Volcano*. US Geological Survey, Professional Paper 96-677, 1996.

Global Volcanism Program. "Report on Ontakesan (Japan)." In: R. Wun-
derman, ed., *Bulletin of the Global Volcanism Network* 40 (Smithsonian Institution, 2015): 3.

Grove, T. L., C. B. Till, and M. J. Krawczynski. "The Role of H_2O in
Subduction Zone Magmatism." *Annual Review Earth and Planetary
Sciences* 40 (2012): 413–39.

Gudmundsson, Magnús T., Kristín Jónsdóttir, Andrew Hooper, et al.
"Gradual Caldera Collapse at Bárdarbunga Volcano, Iceland, Regulated by Lateral Magma Outflow." *Science* 353, no. 6296 (2016):
aaf8988. https://doi.org/10.1126/science.aaf8988

Hansen, S. M., B. Schmandt, A. Levander, et al. "Seismic Evidence for a Cold Serpentinized Mantle Wedge beneath Mount St Helens." *Nature Communications* 7, no. 13242 (2016). https://doi.org/10.1038/ncomms13242

Hawaii Center for Volcanology. "Loihi Volcano." https://www.soest.hawaii.edu/GG/HCV/loihi.html

How Volcanoes Work. "Nevado Del Ruiz (1985)." http://www.geology.sdsu.edu/how_volcanoes_work/Nevado.html

How Volcanoes Work. http://www.geology.sdsu.edu/how_volcanoes_work/index.html

Huang, Hsin-Hua, Fan-Chi Lin, Brandon Schmandt, et al. "The Yellowstone Magmatic System from the Mantle Plume to the Upper Crust." *Science* 348, no. 6236 (2015): 773–76. https://doi.org/10.1126/science.aaa5648

Iceland (website). "Volcanoes." http://www.iceland.is/the-big-picture/nature-environment/volcanoes

Live Science. "The Science behind Hawaii's 'Smiley Face' Volcano." https://www.livescience.com/55630-science-of-Kilauea -volcano -smiley-face.html

Live Science. "What Would Happen If Yellowstone's Supervolcano Erupted?" https://www.livescience.com/20714-yellowstone -supervolcano-eruption.html

Loomis, I. "Faster Lava Flows, Explosive Eruptions Begin at Kilauea." *Eos* 99 (2018). https://doi.org/10.1029/2018EO099655

Lowenstern, J. B., T. W. Sisson, and S. Hurwitz. "Probing Magma Reservoirs to Improve Volcano Forecasts." *Eos* 98 (2017). https://doi.org/10.1029/2017EO085189

Major, Jon, Richard Waitt, Alexa Van Easton, et al. "Mount St. Helens: The 1980 and Later Eruptions and Effects in Toutle River Valley." IAVCEI 2017 Scientific Assembly Mid-Conference Field Trip Guide, 2017.

Manga, Michael, Simon A. Carn, Katharine V. Cashman, et al. *Volcanic Eruptions and Their Repose, Unrest, Precursors, and Timing.* Washington, DC: National Academies Press, 2017.

Mastin, L. G., A. R. Van Eaton, and J. B. Lowenstern. "Modeling Ash Fall Distribution from a Yellowstone Supereruption." *Geochemistry, Geophysics, Geosystems* 15 (2014): 3459–75. https://doi.org/10.1002/2014GC005469

Mileti, D. S., P. A. Bolton, G. Fernandez, and R. G. Updike. *The Eruption of Nevado Del Ruiz Volcano, Colombia, South America November 13, 1985.* Washington, DC: National Academies Press, 1991.

Mount St. Helens Information Resource Center. "History: Mount St. Helens." http://www.mountsthelens.com/history-1.html

National Geographic. "Q&A: Why Iceland's Volcanoes Have Vexed Humans for Centuries." http://news.nationalgeographic.com/news /2014/08/140822-iceland-volcano-eruption-bardarbunga-ice -witze/

National Park Service. "Visitation to Hawai'i Volcanoes National Park in 2015 Creates $151,246,200 in Economic Benefits." https://www.nps .gov/havo/learn/news/20160428_pr_visitation.htm

Neal, C. A., S. R. Brantley, L. Antolik, et. al. "The 2018 Rift Eruption and Summit Collapse of Kīlauea Volcano." *Science* 567, no. 6425 (2019): 367–74. https//doi.org/10.1126/science.aav7046

Nelson, Peter L., and Stephen P. Grand. "Lower-Mantle Plume beneath the Yellowstone Hotspot Revealed by Core Waves." *Nature Geoscience* 11 (2018): 280–84. https://doi.org/10.1038/s41561-018-0075-y

New York Times. "An Icelandic Volcano Reveals Secrets of Its Eruption." https://www.nytimes.com/2016/07/15/science/bardarbunga -volcano-caldera-collapse.html

New York Times. "Surprise from the Supervolcano under Yellowstone." https://www.nytimes.com/2017/10/10/science/yellowstone -volcano-eruption.html

Newhall, Christopher G., and Raymondo S. Punongbayan, eds. *Fire and Mud: Eruptions and Lahars of Mount Pinatubo, Philippines.* Seattle: University of Washington Press, 1996. https://pubs.usgs.gov /pinatubo/prelim.html.

Nooner, S. L., and W. W. Chadwick Jr. "Inflation-Predictable Behavior and Co-Eruption Deformation at Axial Seamount." *Science* 354, no. 6318 (2016): 1399–1403. https://doi.org/10.1126/science.aah4666

Paris, R., J. J. Coello Bravo, M. E. M. Gonzalez, et al. "Explosive Eruption, Flank Collapse and Megatsunami at Tenerife ca. 170 ka." *Nature Communications* 8 (2017): 15246. https://doi.org/10.1038 /ncomms15246

Patrick M. R., K. R. Anderson, M. P. Poland, et al. "Lava Lake Level as a Gauge of Magma Reservoir Pressure and Eruptive Hazard." *Geology* 43 (2015): 831–34. https://doi.org/10.1130/G36896.1

Patrick M. R., T. Orr, D. A. Swanson, and E. Lev. "Shallow and Deep Controls on Lava Lake Surface Motion at Kīlauea Volcano." *Journal of Volcanology and Geothermal Research* 328 (2016): 247–61. http://doi .org/10.1016/j.jvolgeores.2016.11.010

Patrick M. R., T. Orr, G. Fisher, et al. "Thermal Mapping of a Pahoehoe Lava Flow, Kīlauea Volcano." *Journal of Volcanology and Geothermal*

Research 332 (2016): 71–87. http://dx.doi.org/10.1016/j.jvolgeores
.2016.12.007

Patrick, Matthew R., Donald Swanson, and Tim Orr. "Automated Tracking of Lava Lake Level Using Thermal Images at Kīlauea Volcano,
Hawai'i." *Journal of Applied Volcanology Society and Volcanoes* 5
(2016): 6. https://doi.org/10.1186/s13617-016-0047-0

Poland, M. P., T. J. Takahashi, and C. M. Landowski, eds. *Characteristics
of Hawaiian Volcanoes*. US Geological Survey, Professional Paper
1801, 2014. http://dx.doi.org/10.3133/pp1801.

Poland, Michael P., Asta Miklius, A. Jeff Sutton, and Carl R. Thornber.
"A Mantle-Driven Surge in Magma Supply to Kīlauea Volcano during
2003–2007." *Nature Geoscience* 5 (2012): 295–300. https://doi.org
/10.1038/ngeo1426

Prager, Ellen. *Furious Earth: The Science and Nature of Earthquakes, Volcanoes, and Tsunamis*. New York: McGraw-Hill, 2000.

Ramalho, Ricardo S., Gisela Winckler, José Madeira, et al. "Hazard
Potential of Volcanic Flank Collapses Raised by New Megatsunami
Evidence." *Science Advances* 1, no. 9 (2015): e1500456. https://doi
.org/10.1126/sciadv.1500456

Rougier, J., S. Sparks, K. Cashman, and S. Brown. "The Global
Magnitude-Frequency Relationship for Large Explosive Volcanic
Eruptions." *Earth and Planetary Science Letters* 482 (2018). https://
doi.org/10.1016/j.epsl.2017.11.015

Scientific American. "Excitement on Etna: The Danger of Phreatic
Eruptions." https://blogs.scientificamerican.com/rosetta-stones
/excitement-on-etna-the-danger-of-phreatic-eruptions-video/

Sigmundsson, F. "New Insights into Magma Plumbing along Rift
Systems from Detailed Observations of Eruptive Behavior at Axial
Volcano." *Geophysical Research Letters* 43, no. 24 (2016): 12,423–27.
https://doi.org/10.1002/2016GL071884

Singer, Brad S., Hélène Le Mével, Joseph M. Licciardi, et al. "Geomor-
phic Expression of Rapid Holocene Silicic Magma Reservoir Growth
beneath Laguna del Maule, Chile." *Science Advances* 4, no. 6 (2018):
eaat1513. https://doi.org/10.1126/sciadv.aat1513

Singer, Brad S., Nathan L. Andersen, Hélène Le Mével, et al. 2014. "Dynamics of a Large, Restless, Rhyolitic Magma System at Laguna del
Maule, Southern Andes, Chile." *GSA Today* 24, no. 12 (2014): 4–10.
https://doi.org/10.1130/GSATG216A.1

Smithsonian Institution Global Volcanism Program. "USGS Weekly Volcanic Activity Report." http://volcano.si.edu/reports_weekly.cfm

Sparks, R. S. J., S. Self, J. Gratten, et al. *Supereruptions: Global Effects* and *Future Threats*. Report of a Geological Society Working Group. Geological Society of London, 2005.

Swanson, Donald A., Timothy R. Rose, Adonara E. Mucek, et al. "Cycles of Explosive and Effusive Eruptions at Kīlauea Volcano, Hawai'i." *Geology* 42, no. 7 (2014): 631–34. https://doi.org/10.1130/G35701.1

Thelen, W. A., A. Miklius, and C. Neal. "Volcanic Unrest at Mauna Loa, Earth's Largest Active Volcano." *Eos* 98 (2017). https://doi.org/10.1029/2017EO083937

Thompson, Dick. *Volcano Cowboys: The Rocky Evolution of a Dangerous Science*. New York: St. Martin's Press, 2000.

Thordarson, T., and G. Larsen. "Volcanism in Iceland in Historical Time: Volcano Types, Eruption Styles and Eruptive History." *Journal of Geodynamics* 43, no. 1 (2007): 118–52. https://doi.org/10.1016/j.jog.2006.09.005

Thordarson, T., and S. Self. "Atmospheric and Environmental Effects of the 1783–1784 Laki Eruption: A Review and Reassessment." *Journal of Geophysical Research* 108, no. D1 (2003): AAC 7-1–AAC 7-29. https://doi.org/10.1029/2001JD002042

Thordarson, T., and S. Self. "The Laki (Skaftár Fires) and Grimsvötn Eruptions in 1783–1785." *Bulletin of Volcanology* 55 (1993): 233–63.

Tilliing, R. I., C. Heliker, and D. A. Swanson. *Eruptions of Hawaiian Volcanoes—Past, Present, and Future*. US Geological Survey, General Information Product 117, 2010. https://pubs.usgs.gov/gip/117/

Tilling, Robert. "Mount St. Helens 20 Years Later: What We've Learned." *Geotimes* (May 2000). http://www.geotimes.org/may00/featurestory.html

United States Geological Survey. "Kīlauea: An Explosive Volcano in Hawaii." USGS Fact Sheet 2011-3064, 2011.

United States Geological Survey. "Loihi." https://volcanoes.usgs.gov/volcanoes/loihi/

United States Geological Survey. "Mount St. Helens, 1980 to Now—What's Going On?" USGS Fact Sheet 2013-3014, 2013.

United States Geological Survey. "Mount St. Helens: From the 1980 Eruption to 2000." https://pubs.usgs.gov/fs/2000/fs036-00/

United States Geological Survey. "Questions about Yellowstone Volcanic History." https://volcanoes.usgs.gov/volcanoes/yellowstone/yellowstone_sub_page_54.html

United States Geological Survey. "Study to Uncover Yellowstone's Subsurface Mysteries." https://www.usgs.gov/news/study-uncover-yellowstone-s-subsurface-mysteries

United States Geological Survey. "The Caldera Chronicles: The Source

of Yellowstone's Heat." https://volcanoes.usgs.gov/volcanoes
/yellowstone/article_home.html?vaid=22

United States Geological Survey. "The Cataclysmic 1991 Eruption of
Mount Pinatubo, Philippines." https://pubs.usgs.gov/fs/1997/fs113
-97/fs113-97.pdf

United States Geological Survey. "The Climactic Eruption of May 18,
1980." https://pubs.usgs.gov/gip/msh/climactic.html

University of Washington. "Geophysicists Prep for Massive 'Ultrasound'
of Mount St. Helens. http://www.washington.edu/news/2014/07
/17/geophysicists-prep-for-massive-ultrasound-of-mount-st-helens/

University of Wisconsin–Madison. "UW Team Explores Large, Restless
Volcanic Field in Chile." https://news.wisc.edu/uw-team-explores
-large-restless-volcanic-field-in-chile/

Voight, B. "The 1985 "Nevado del Ruiz Volcano Catastrophe: Anatomy
and Restrospection." *Journal of Volcanology and Geothermal Research*
44 (1990): 349–86.

Volcano Discovery. "Holuhraun Fissure Eruption 2014-15 at Bardar-
bunga Volcano, Iceland." https://www.volcanodiscovery.com
/bardarbunga/seismic-crisis-2014/updates.html

Waitt, R. *Path of Destruction: Eyewitness Chronicles of Mount St. Helens.*
Washington: Washington State University Press, 2014.

Wendel, J., and M. Kumar. "Pinatubo 25 Years Later: Eight Ways the
Eruption Broke Ground." *Eos* 97 (2016). https://doi.org/10.1029
/2016EO053889

Wilcock, W. S. D., M. Tolstoy, F. Waldhauser, et al. "Seismic Constraints
on Caldera Dynamics from the 2015 Axial Seamount Eruption."
Science 354, no. 6318 (2016): 1395–99. https://doi.org/10.1126
/science.aah5563

Wired. "A Caldera in the Making? The Curious Story of Laguna Del
Maule." https://www.wired.com/2013/11/a-caldera-eruption-in-the
-making-the-curious-story-of-laguna-del-maule/

Wired. "Local and Global Impacts of the 1783–84 Laki Eruption."
https://www.wired.com/2013/06/local-and-global-impacts-1793
-laki-eruption-iceland/

Wired. "Over 30 Hikers Die during Ontake Eruption in Japan: What
Happened?" https://www.wired.com/2014/09/30-hikers-die-ontake
-eruption-japan-happened/

Witze, Alexandra, and Jeff Kanipe. *Island on Fire.* London: Profile, 2014.

World Volcano Observatories. http://www.wovo.org/homepage-wovo
.org/

Wright, T. L., and F. W. Klein. "Two Hundred Years of Magma Transport
and Storage at Kīlauea Volcano Hawai'i, 1790–2008." US Geologi-

cal Survey Professional Paper 1806 (2014). http://doi.org/10.3133
/pp1806

Earthquakes and Tsunamis

Ágústsdóttir, T., J. Woods, T. Greenfield, et al. "Strike-Slip Faulting
during the 2014 Bárðarbunga-Holuhraun Dike Intrusion, Central
Iceland." *Geophysical Research Letters* 43 (2016): 1495–1503. https://
doi.org/10.1002/2015GL067423

Arcos, N., P. Dunbar, K. Stroker, and L. Kong. "The Legacy of the 1992
Nicaragua Tsunami." *Eos* 98 (2017). https://doi.org/10.1029
/2017EO080845

Atwater, B. F. "Evidence for Great Holocene Earthquakes along the Outer
Coast of Washington State." *Science* 236 (1987): 942–44.

Bendick, R., and R. Bilham. "Do Weak Global Stresses Synchronize
Earthquakes?" *Geophysical Research Letters* 44 (2017): 8320–27.
https//doi.org/10.1002/2017GL074934

Benz, H. M., R. L. Dart, Antonio Villaseñor, et al. "Seismicity of the
Earth 1900–2010 Mexico and Vicinity." US Geological Survey Open-
File Report 2010–1083-F (2011). https://pubs.usgs.gov/of/2010
/1083/f/

Bird, P., D. D. Jackson, Y. Y. Kagan, et al. "GEAR1: A Global Earthquake
Activity Rate Model Constructed from Geodetic Strain Rates and
Smoothed Seismicity." *Bulletin of the Seismological Society of America*
105, no. 5 (2015): 2538–54. https://doi.org/10.1785/0120150058

Bletery, Q., A. M. Thomas, A. W. Rempel, et al. "Mega-Earthquakes Rupture Flat Megathrusts." *Science* 354, no. 6315 (2016): 1027. https//
doi:10.1126/science.aag0482

Bo Li, Abhijit Ghosh. "Near-Continuous Tremor and Low-Frequency
Earthquake Activities in the Alaska-Aleutian Subduction Zone Revealed by a Mini Seismic Array." *Geophysical Research Letters* 44, no.
11 (2017): 5427. https://doi.org/10.1002/2016GL072088

Bolt, Bruce A. 1993. *Earthquakes and Geological Discovery*. New York:
Scientific American Library, 1993.

California Institute of Technology. "Rethinking the Causes of Giant
Earthquakes." http://www.tectonics.caltech.edu/outreach/highlights
/sumatra/rethink.html

Dixon, Timothy H. *Curbing Catastrophe: Natural Hazards and Risk Reduction in the Modern World*. Cambridge: Cambridge University Press,
2017.

Dixon, Timothy H., Yan Jiang, Rocco Malservisi, et al. "Earthquake
and Tsunami Forecasts: Relation of Slow Slip Events to Subsequent

Earthquake Rupture." *Proceedings of the National Academy of Sciences* 111, no. 48 (2014): 17039–44. https://doi.org/10.1073/pnas .1412299111

Dragert, Herb, Kelin Wang, and Thomas S. James. "A Silent Slip Event on the Deeper Cascadia Subduction Interface." *Science* 292, no. 5521 (2001): 1525–28. https://doi.org./10.1126/science.1060152

Dvorak, John. *Earthquake Storms: The Fascinating History and Volatile Future of the Sand Andreas Fault*. New York: Pegasus, 2014.

Earthscope. "Locked, Loaded and Looming?" http://www.earthscope.org /articles/SN_smith_konter

Earthscope. "Tracking the Fluids in a Weak Fault." http://www .earthscope.org/articles/Tracking_the_fluids

Ellsworth, W. L. "Injection-Induced Earthquakes." *Science* 341, no. 6142 (2013). https://doi.org/10.1126/science.1225942

Fujiwara1, Toshiya, Shuichi Kodaira, Tetsuo No, et al. "The 2011 Tohoku-Oki Earthquake: Displacement Reaching the Trench Axis." *Science* 334, no. 6060 (2011): 1240. https//doi:10.1126/science .1211554

Geist, Eric L., Vasily V. Titov, and Costas E. Synolakis. "Tsunami: Wave of Change." *Scientific American* 294, no. 1 (200663): 57–63.

Global Earthquake Model. https://www.globalquakemodel.org

Goldfinger, C., C. H. Nelson, A. E. Morey, et al. "Turbidite Event History: Methods and Implications for Holocene Paleoseismicity of the Cascadia Subduction Zone." US Geological Survey Professional Paper 1661–F (2012). https://pubs.usgs.gov/pp/pp1661f/

Goldfinger, C., Y. Ikeda, R. S. Yeats, and J. Ren. "Superquakes and Supercycles." *Seismological Research Letters* 84, no. 1 (2013): 24. https//doi .org/10.1785/0220110135

Goldfinger, Chris, Kelly Grijalva, Roland Bürgmann, et al. 2008. "Late Holocene Rupture of the Northern San Andreas Fault and Possible Stress Linkage to the Cascadia Subduction Zone." *Bulletin of the Seismological Society of America* 98, no. 2 (2008): 861–89. https://doi .org/10.1785/0120060411

Goldfinger, Chris, Steve Galera, Jeffery Beesona, et al. "The Importance of Site Selection, Sediment Supply, and Hydrodynamics: A Case Study of Submarine Paleoseismology on the Northern Cascadia Margin, Washington USA." *Marine Geology* 384 (2017): 4–46. https://doi .org/10.1016/j.margeo.2016.06.008

Gomberg, Joan. "Slow-Slip Phenomena in Cascadia from 2007 and Beyond: A Review." *Geological Society of America Bulletin* 122, nos. 7/8 (2010): 963–78. https//doi.org/10.1130/B30287.1

Great California Shake Out. https://www.shakeout.org/california/

Hamling, Ian J., Sigrun Hreinsdóttir, Kate Clark, et al. "Complex Multi-fault Rupture during the 2016 Mw 7.8 Kaikoura Earthquake, New Zealand." *Science* 356, no 6334 (2017): eaam7194. https://doi.org/10.1126/science.aam7194

Hauksson, Egill, Lucile M. Jones, Kate Hutton, and Donna Eberhart-Phillips. "The 1992 Landers Earthquake Sequence: Seismological Observations." *Journal Geophysical Research* 98, no. B11 (1993): 19,835–58

Hough, Susan Elizabeth. *Earthshaking Science: What We Know (and Don't Know) about Earthquakes*. Princeton, NJ: Princeton University Press, 2002.

Hough, Susan. *Predicting the Unpredictable: The Tumultuous Science of Earthquake Prediction*. Princeton, NJ: Princeton University Press, 2010.

Kanamori, Hiroo, and Masayuki Kikuchi "The 1992 Nicaragua Earthquake: A Slow Tsunami Earthquake Associated with Subducted Sediments." *Nature* 361 (1993): 714–16. https://doi.org/10.1038/361714a0

Kerr, Richard A. "A Tantalizing View of What Set Off Japan's Killer Quake." *Science* 335, no. 6066 (2012): 272. https//doi:10.1126/science.335.6066.272

Lawson, A. C. "The California Earth-quake of April 18, 1906: Report of the State Earthquake Investigation Commission." *Carnegie Institution of Washington Publication* 87, no. 1 (1908). http://bancroft.berkeley.edu/collections/earthquakeandfire/splash.html

Levin, B. W., E. V. Sasarova, G. M. Steblov, et al. "Variations of the Earth's Rotation Rate and Cyclic Processes in Geodynamics." *Geodesy and Geodynamics* 8, no. 3 (2017): 206–12. https://doi.org/10.1016/j.geog.2017.03.007

Litchfield, Nicola J., Pilar Villamor, Russ J. Van Dissen, et al. "Surface Rupture of Multiple Crustal Faults in the 2016 Mw7.8 Kaikoura, New Zealand, Earthquake." *Bulletin of Seismological Society of America* 108, 3B (2018): 1496-1520. https://doi.org/10.1785/0120170300

Live Science. "Earthquake Watch: California Is Overdue for a 'Big One'." https://www.livescience.com/61601-california-overdue-earthquake.html

Live Science. "Great Mysteries: What Happens Inside an Earthquake." https://www.livescience.com/1810-greatest-mysteries-earthquake.html

Live Science. "Landslide-Driven Megatsunamis Threaten Hawaii." https://www.livescience.com/25293-hawaii-giant-tsunami-landslides.html

Live Science. "Slippery Clays at Fault." https://www.livescience.com
/34677-how-slow-earthquakes-work.html

Live Science. "The Science behind Japan's Deadly Earthquake." https://
www.livescience.com/13177-japan-deadly-earthquake-tsunami.html

Los Angeles Times. "East Bay Fault Is 'Tectonic Time Bomb,' More Dangerous Than San Andreas, New Study Finds." http://www.latimes
.com/local/lanow/la-me-hayward-fault-20180417-htmlstory.html

Lozos, Julian C. "A Case for Historic Joint Rupture of the San Andreas
and San Jacinto Faults." *Science Advances* 2, no. 3 (2016): e1500621.
https://doi.org/10.1126/sciadv.1500621

Mapping Ignorance. "What's Going On beneath Mexico?" https://
mappingignorance.org/2017/09/22/whats-going-beneath-mexico/

McCaffrey, Robert. "Global Frequency of Magnitude 9 Earthquakes." *Geology* 36, no. 3 (2008): 263–66. https://doi.org/10.1130/G24402A.1

McCaffrey, Robert. "The Next Great Earthquake." *Science* 315, no. 5819
(2007): 1675–76. https:/doi.org/10.1126/science.1140173

Meier, M. A., J. P. Ampuero, and T. H. Heaton. "The Hidden Simplicity of
Subduction Megathrust Earthquakes." *Science* 357, no. 6357 (2017):
1277–81. https//doi.org/10.1126/science.aan5643

Miles, Kathryn. *Quakeland: On the Road to America's Next Devastating
Earthquake.* New York: Dutton, 2017.

Muffler, B. *Images of America: Hawaii Tsunamis.* Charleston, SC: Arcadia,
2015.

National Oceanic and Atmospheric Administration. "Tsunami Historical Series: Aleutian Islands—1946." https://sos.noaa.gov/datasets
/tsunami-historical-series-aleutian-islands-1946/

National Oceanic and Atmospheric Administration. "Tsunami Warning
Center History." http://www.tsunami.gov/?page=history

National Science Foundation. "When Nature Strikes: Science of Natural
Hazards." https://www.nsf.gov/news/special_reports/naturestrikes
/index.jsp

New York Times. "Scientists in Mexico Scramble to Deploy Seismic
Sensors." https://www.nytimes.com/2017/10/02/science/mexico
-earthquakes-prediction.html

New York Times. "What Science Can, and Can't Do, to Cut Losses from
the Next Tsunami." https://dotearth.blogs.nytimes.com/2014/12
/26/what-science-can-and-cant-do-to-cut-losses-from-the-next
-tsunami/?_r=0

Normile, Dennis. "Scientific Consensus on Great Quake Came Too
Late." *Science* 332, no. 6025 (2011): 22–23. https://doi.org/10.1126
/science.332.6025.22

Obara, Kazushige, and Aitaro Kato. "Connecting Slow Earthquakes to

Huge Earthquakes." *Science* 353, no. 6296 (2016): 253–57. https://doi.org/DOI: 10.1126/science.aaf1512

Obara, Kazushige. "Nonvolcanic Deep Tremor Associated with Subduction in Southwest Japan." *Science* 296, no. 5573 (2002): 1679–81. https://doi.org/10.1126/science.1070378

Oklahoma Geological Survey. "Statement on Oklahoma Seismicity April 21, 2015." http://wichita.ogs.ou.edu/documents/OGS_Statement-Earthquakes-4-21-15.pdf

Oregon State University. "Scientists Possible Link between Cascadia Zone, San Andreas Fault." http://oregonstate.edu/ua/ncs/archives/2008/apr/scientists-eye-possible-link-between-cascadia-zone-san-andreas-fault

Popular Science. "Earthquakes Are Even Harder to Predict Than We Thought." https://www.popsci.com/earthquake-harder-to-predict-than-we-thought

Radiguet, M., H. Perfettini, N. Cotte, et al. "Triggering of the 2014 Mw7.3 Papanoa Earthquake by a Slow Slip Event in Guerrero, Mexico." *Nature Geoscience* 9 (2016): 829–33. https://doi.org/10.1038/ngeo2817

Reid, H. F. "The Mechanics of the Earthquake: The California Earthquake of April 18, 1906." Report of the State Investigation Commission 2. Washington, DC: Carnegie Institution of Washington, 1910.

Rogers, Gary, and Herb Dragert. "Episodic Tremor and Slip on the Cascadia Subduction Zone: The Chatter of Silent Slip." *Science* 300, no. 5627 (2003): 1942–43. https://doi.org/10.1126/science.1084783

Rosen, Julia. "Seismic Array Shifts to Alaska." *Science* 358, no. 6359 (2017): 22. https://10.1126/science.358.6359.22

Sahakian, Valerie, Annie Kell, Alistair Harding, et al. "Geophysical Evidence for a San Andreas Subparallel Transtensional Fault along the Northeastern Shore of the Salton Sea." *Bulletin of the Seismological Society of America* 106, no. 5 (2016): 1963–78. https://doi.org/10.1785/0120150350

Satake, K., K. Wang, and B. F. Atwater. "Fault Slip and Seismic Moment of the 1700 Cascadia Earthquake Inferred from Japanese Tsunami Descriptions." *Journal Geophysical Research* 108 (2003): 2325.

Satake, Kenji, Kunihiko Shimazaki, Yoshinobu Tsuji, and Kazue Ueda. "Time and Size of a Giant Earthquake in Cascadia Inferred from Japanese Tsunami Records of January 1700." *Nature* 379 (1996): 246–49. https//doi:10.1038/379246a0

Sato, M., T. Ishikawa, N. Ujihara, et al. "Displacement above the Hypocenter of the 2011 Tohoku–Oki Earthquake." *Science* 332 (2011): 1395.

Schellart, W. P., and N. Rawlison. "Global Correlations between Maximum Magnitudes of Subduction Zone Interface Thrust Earthquakes and Physical Parameters of Subduction Zones." *Physics of the Earth and Planetary Interiors* 225 (2013): 41–67. https://doi.org/10.1016/j.pepi.2013.10.001

Science. "Sloshing of Earth's Core May Spike Major Earthquakes." http://www.sciencemag.org/news/2017/10/sloshing-earth-s-core-may-spike-major-earthquakes

Science. "Unusual Mexico Earthquake May Have Relieved Stress in Seismic Gap." http://www.sciencemag.org/news/2017/09/unusual-mexico-earthquake-may-have-relieved-stress-seismic-gap

Scientific American. "Why the Mexico City Earthquake Shook Up Disaster Predictions." https://www.scientificamerican.com/article/why-the-mexico-city-earthquake-shook-up-disaster-predictions1/

Smithsonian. "Slow Earthquakes Are a Thing." https://www.smithsonianmag.com/science-nature/slow-earthquakes-are-thing-180960248/

Stanford University. "Advancing Earthquake and Tsunami Science: Tohoku Four Years Later." https://pangea.stanford.edu/news/advancing-earthquake-and-tsunami-science-tōhoku-four-years-later

State of Oregon. "The Oregon Resilience Plan—Cascadia: Oregon's Greatest Natural Threat." February 2013. http://www.oregon.gov/oem/Documents/01_ORP_Cascadia.pdf

Taira, T., R. Bürgmann, R. M. Nadeau, and D. S. Dreger. "Variability of Fault Slip Behavior along the San Andreas Fault in the San Juan Bautista Region." *Journal Geophysical Research* 119 (2014): 8827–44. https//doi.org/:10.1002/2014JB011427

Temblor. "California Bill to Create a Public Inventory of Collapse-Risk Buildings Passes Assembly, Heads for Senate." http://temblor.net/earthquake-insights/new-california-bill-aims-to-create-a-public-inventory-of-collapse-risk-buildings-6497/

Temblor. "Can Changes in Earth's Rotation Be Used to Forecast Earthquakes?" http://temblor.net/earthquake-insights/can-changes-in-earths-rotation-be-used-to-forecast-earthquakes-5642/

Temblor. "Creating the Ties That Bind: Breaking Bread and Sharing Stories in Latin America." http://temblor.net/earthquake-insights/creating-the-ties-that-bind-breaking-bread-and-sharing-stories-in-latin-america-6661/

Temblor. "M=6.1 Mexican Aftershock Strongly Promoted by M=8.1 Chiapas Mainshock." http://temblor.net/earthquake-insights

/m6-1-mexican-aftershock-strongly-promoted-by-m8-1-chiapas
-mainshock-5317/

Temblor. "Mexican Earthquakes: Chain Reaction or Coincidence?"
http://temblor.net/earthquake-insights/chiapas-and-puebla-mexico
-earthquakes-chain-reaction-or-coincidence-5248/

Temblor. "Where the San Andreas Goes to Get Away from It All." http://
temblor.net/earthquake-insights/san-andreas-fault-on-ice-5599/

United States Geological Survey. "1906 Marked the Dawn of the Scientific Revolution." https://earthquake.usgs.gov/earthquakes/events
/1906calif/18april/revolution.php

United States Geological Survey. "Earthquake Statistics." https://
earthquake.usgs.gov/earthquakes/browse/stats.php

United States Geological Survey. "Tsunami Generation from the 2004
M=9.1 Sumatra-Andaman Earthquake." https://walrus.wr.usgs.gov
/tsunami/sumatraEQ/model.html

United States Geological Survey. "UCERF3: A New Earthquake Forecast
for California's Complex Fault System." https://pubs.usgs.gov/fs
/2015/3009/pdf/fs2015-3009.pdf

USArray. http://www.usarray.org

Vidale, John, and Heidi Houston. "Slow Slip: A New Kind of Earthquake." *Physics Today* 65 (2012): 38–43. https://doi.org/10.1063
/PT.3.1399

Wang, Kelin, and Masataka Kinoshita. "Dangers of Being Thin and
Weak." *Science* 342, no. 6163 (2013): 1178–80. https//doi.org/10
.1126/science.1246518

Western States Seismic Policy Council. "1946 Aleutians Tsunami."
http://www.wsspc.org/resources-reports/tsunami-center/significant
-tsunami-events/1946-aleutians-tsunami/

Woods Hole Oceanographic Institution. "Lessons from the 2011 Japan
Quake." http://www.whoi.edu/oceanus/feature/lessons-from-the
-2011-japan-quake

Yeats, Robert S. *Earthquake Time Bombs*. Cambridge: Cambridge University Press, 2015.

Yeats, Robert S. *Living with Earthquakes in the Pacific Northwest*. Corval-
lis: Oregon State University Press, 2004.

Zilio, Luca Dal, Ylona van Dinther, Tara V. Gerya, and Casper C. Pranger.
"Seismic Behaviour of Mountain Belts Controlled by Plate Convergence Rate." *Earth and Planetary Science Letters* 482 (2018): 81–92.
https://doi.org/10.1016/j.epsl.2017.10.053

Zoback, Mary Lou. "The 1906 Earthquake and a Century of Progress in
Understanding Earthquakes and Their Hazards." *Geological Society*

of America Today 16, nos. 4/5 (2006): 4–11. https://doi.org/10.1130/GSAT01604.1

Hurricanes and Tornados

Air and Space magazine. "Meet Coyote, the Latest (and Smallest) Hurricane Hunter." https://www.airspacemag.com/daily-planet/noaa-sacrifices-drones-appease-poseidon-180965188/

American Broadcasting Company. "Hurricane Irene: Pop-Tarts Top List of Hurricane Purchases." http://abcnews.go.com/US/hurricanes/hurricane-irene-pop-tarts-top-list-hurricane-purchases/story?id=14393602

Bhatia, K. T., G. A. Vecchi, T. K. Knutson, et. al. "Recent Increases in Tropical Cyclone Intensification Rates." *Nature Communications* 10, no. 635 (2019). https://doi.org/10.1038/s41467-019-08471-z

Brinkley, Douglass. *The Great Deluge: Hurricane Katrina, New Orleans, and the Mississippi Gulf Coast*. New York: Harper Perennial, 2006.

Climate Central. "How Global Warming Made Hurricane Sandy Worse." http://www.climatecentral.org/news/how-global-warming-made-hurricane-sandy-worse-15190

Colorado State University. "Summary of 2017 Atlantic Tropical Cyclone Activity and Verification of Authors' Seasonal and Two-Week Forecasts." https://tropical.colostate.edu/media/sites/111/2017/11/2017-11.pdf

Conversation. "Post-Fire Mudslide Problems Aren't New and Likely to Get Worse." https://theconversation.com/post-fire-mudslide-problems-arent-new-and-likely-to-get-worse-90048

Daily News. "Mayor Bloomberg Update on Hurricane Sandy Storm Prep: Don't Be Complacent, NYC." http://www.nydailynews.com/blogs/dailypolitics/mayor-bloomberg-update-hurricane-sandy-storm-prep-don-complacent-nyc-blog-entry-1.1692614

Dispatch. "Ask Rufus: Lt. Col. Duckworth and the 'Surprise Hurricane' of 1943." http://www.cdispatch.com/opinions/article.asp?aid=60542

Emanuel, K. "Will Global Warming Make Hurricane Forecasting More Difficult?" American Meteorological Society online (2016). http:/doi.org/10.1175/BAMS-D-16-0134.1

Falko, J., and S. S. Chen. "Predictability and Dynamics of Tropical Cyclone Rapid Intensification Deduced from High-Resolution Stochastic Ensembles." American Meteorological Society online (2016). https://doi.org/10.1175/MWR-D-15-0413.1

Geotechnical Extreme Events Reconnaissance. http://www.geerassociation.org/reconnaissance-reports/table-view

Greene, C. H., J. A. Francis, and B. C. Monger. "Superstorm Sandy: A Series of Unfortunate Events?" *Oceanography* 26, no. 1 (2013):8–9. https://doi.org/10.5670/oceanog.2013.11

Halverson, Jeffrey B., and Thomas Rabenhorst. "Hurricane Sandy: The Science and Impacts of a Superstorm." *Weatherwise* 66, no. 2 (2013): 14–23. https://doi.org/10.1080/00431672.2013.762838

Hurricane Science. "2004–Hurricane Charley." http://www.hurricane science.org/history/storms/2000s/charley/

Hurricanes Science. "National Hurricane Center Forecast Process." http://www.hurricanescience.org/science/forecast/forecasting /forecastprocess/

Jones, L. *Big Ones: How Natural Disasters Have Shaped Us (and What We Can Do about Them)*. New York: Doubleday, 2018.

Kishore, Nishant, Domingo Marqués, Ayesha Mahmud, et al. "Mortality in Puerto Rico after Hurricane Maria." *New England Journal of Medicine* (May 2018). https://doi.org/10.1056/NEJMsa1803972

Larson, E. *Isaac's Storm*. New York: Vintage, 2000.

Live Science. "How Do Hurricanes Spawn Tornadoes?" https://www .livescience.com/37235-how-hurricanes-spawn-tornadoes.html

National Aeronautics and Space Administration. "Global Hawk." https://www.nasa.gov/centers/armstrong/aircraft/GlobalHawk /index.html

National Oceanic and Atmospheric Administration. "Hurricane Harvey & Its Impacts on Southeast Texas from August 25th to 29th, 2017." http://www.weather.gov/hgx/hurricaneharvey

National Oceanic and Atmospheric Administration. "Hurricane Irma 2017." https://www.weather.gov/tae/Irma2017

National Oceanic and Atmospheric Administration. "Major Hurricane Maria—September 20, 2017." http://www.weather.gov/sju /maria2017

National Oceanic and Atmospheric Administration. "NOAA Kicks Off 2018 with Massive Supercomputer Upgrade." http://www.noaa.gov /media-release/noaa-kicks-off-2018-with-massive-supercomputer -upgrade

National Oceanic and Atmospheric Administration. "Service Assessment: Hurricane Charley, August 9-15, 2004." https://www.weather .gov/media/publications/assessments/Charley06.pdf

National Oceanic and Atmospheric Administration. "Storm Surge Frequently Asked Questions." https://www.nhc.noaa.gov/surge /faq.php

National Oceanic and Atmospheric Administration. "The Saffir-Simpson Hurricane Wind Scale." https://www.nhc.noaa.gov/pdf/sshws.pdf

National Oceanic and Atmospheric Administration. "The State of Hurricane Forecasting." https://noaanhc.wordpress.com/2018/03/09/the-state-of-hurricane-forecasting/

National Oceanic and Atmospheric Administration. "Tropical Cyclone Intensity Forecasting: Still a Challenging Proposition." https://www.nhc.noaa.gov/outreach/presentations/NHC2017_IntensityChallenges.pdf

National Oceanic and Atmospheric Administration. "Tropical Cyclone Report: Hurricane Harvey." https://www.nhc.noaa.gov/data/tcr/AL092017_Harvey.pdf

National Oceanic and Atmospheric Administration. "Tropical Cyclone Report: Hurricane Joaquin." https://www.nhc.noaa.gov/data/tcr/AL112015_Joaquin.pdf

National Oceanic and Atmospheric Administration. "Tropical Cyclone Report: Hurricane Sandy." http://www.nhc.noaa.gov/data/tcr/AL182012_Sandy.pdf

National Oceanic and Atmospheric Administration. "Vertical Datum." https://noaanhc.wordpress.com/tag/vertical-datum/

National Oceanic and Atmospheric Administration. "What Is an Easterly Wave?" http://www.aoml.noaa.gov/hrd/tcfaq/A4.html

National Oceanic and Atmospheric Administration. "When Is Hurricane Season?" http://www.aoml.noaa.gov/hrd/tcfaq/G1.html

Prager, Ellen. *The Oceans*. New York: McGraw-Hill, 2000.

Scientific American. "Hurricane Harvey: Why Is It So Extreme?" https://www.scientificamerican.com/article/hurricane-harvey-why-is-it-so-extreme/

Scotti, R. A. *Sudden Sea: The Great Hurricane of 1938*. New York: Back Bay, 2003.

Sheets, B., and J. Williams. *Hurricane Watch: Forecasting the Deadliest Storms on Earth*. New York: Vintage, 2001.

Slate. "How to Make Sense of Category 4 Hurricane Joaquin's Unknowable Path." http://www.slate.com/blogs/the_slatest/2015/10/01/hurricane_joaquin_forecast_an_unknowable_storm_approaches_the_east_coast.html

Tippett, M. K., and J. E. Cohen. "Tornado Outbreak Variability Follows Taylor's Power Law of Fluctuation Scaling and Increases Dramatically with Severity." *Nature Communications* 7 (2016): 10668. https://doi.org/10.1038/ncomms10668

Trenberth, K. E., L. Cheng, P. Jacobs, et al. "Hurricane Harvey Links to Ocean Heat Content and Climate Change Adaptation." *Earth's Future* 6, no. 5 (2018): 730–44. https://doi.org/10.1029/2018EF000825

University Corporation for Atmospheric Research. "A More Perfect

Storm: Sandy Could Make U.S. History." https://www2.ucar.edu
/atmosnews/perspective/8192/more-perfect-storm

USA Today. "Supercharged by Global Warming, Record Hot Seawater Fueled Hurricane Harvey." https://www.usatoday.com/story
/news/2018/05/14/hurricane-harvey-record-hot-seawater-global
-warming/607715002/

Weather Underground. "Prepare for a Storm Surge." https://www
.wunderground.com/prepare/storm-surge

Wired. "Harvey Wrecks up to a Million Cars in Car-Dependent Houston."
https://www.wired.com/story/harvey-houston-cars-ruined/

Rogue Waves, Landslides, Rip Currents, Sinkholes, Sharks

British Broadcasting Company. "Terrifying 20m-Tall 'Rogue Waves'
Are Actually Real." http://www.bbc.com/earth/story/20170510
-terrifying-20m-tall-rogue-waves-are-actually-real

Castelle, B., T. Scott, R. W. Brander, and R. J. McCarroll. "Rip Current
Types, Circulation and Hazard." *Earth-Science Reviews* 163 (2016):
1–21. https://doi.org/10.1016/j.earscirev.2016.09.008

Dai, F. C., C. Xu, X. Yao, et al. "Spatial Distribution of Landslides Triggered by the 2008 Ms 8.0 Wenchuan Earthquake, China." *Journal
of Asian Earth Sciences* 40, no. 4 (2011): 883–95. https://doi.org/10
.1016/j.jseaes.2010.04.010

Donelan, Mark A., and Anne-Karin Magnusson. "The Making of the An-
drea Wave and Other Rogues." *Scientific Reports* 7, no. 44124 (2017).
https://doi.org/10.1038/srep44124

Fan, R. L., L. M. Zhang, H. J. Wang, and X. M. Fan. "Evolution Of Debris
Flow Activities In Gaojiagou Ravine During 2008–2016 After The
Wenchuan Earthquake." *Engineering Geology* 235 (2018): 1–10. doi:
https://doi.org/10.1016/j.enggeo.2018.01.017

Gariano, Stefano Luigi, and Fausto Guzzetti. "Landslides in a Changing
Climate." *Earth-Science Reviews* 162 (2016): 227–52. https://doi.org
/10.1016/j.earscirev.2016.08.011

Gemmrich, J., and J. Thomson. "Observations of the Shape and Group
Dynamics of Rogue Waves." *Geophysical Research Letters* 44 (2017):
1823–30. https://doi.org/10.1002/2016GL072398

GIS Lounge. "A Global Landslide Map That Updates Every 30 Minutes."
https://www.gislounge.com/global-landslide-potential-map-updates
-every-30-minutes/

Guardian. "Florida's Most Famous Sinkhole." https://www.theguardian
.com/world/2013/aug/14/florida-most-famous-sinkhole

Highland, L. M., and Peter Bobrowsky. "The Landslide Handbook—

A Guide to Understanding Landslides." US Geological Survey Circular 1325 (2008).

Landslide Blog. "Landslide Weirdness: A Landslide as a 'Crashed UFO.'" https://blogs.agu.org/landslideblog/2018/03/05/landslide-weirdness-ufo/

Leatherman, S. P. "Rip Current Measurements at Three South Florida Beaches." *Journal of Coastal Research* 33, no 5 (2017): 1228–34.

Leatherman, S. P. "Rip Currents in South Florida: A Major Coastal Hazard and Management Challenge." *Journal Coastal Zone Management* 19 (2016): 431. https://doi.org/10.4172/2473-3350.1000431

Leatherman, S. P. "Rip Currents." In *Coastal Hazards*. Springer, 2012.

Live Science. "New Method Predicts Monster Waves." https://www.livescience.com/232-method-predicts-monster-waves.html

Live Science. "Rip Currents: The Ocean's Deadliest Trick." https://www.livescience.com/3910-rip-currents-ocean-deadliest-trick.html

Live Science. "What Are Sinkholes?" https://www.livescience.com/44123-what-are-sinkholes.html

Los Angeles Times. "How a Group of Scientists Are Using the Deadly Montecito Mudflow to Predict Future Disasters." http://www.latimes.com/local/lanow/la-me-ln-montecito-mudflow-usgs-models-20180207-story.html

Los Angeles Times. "Two Weeks Later, Montecito's Devastated Neighborhoods Are Still Empty and Eerily Quiet." http://www.latimes.com/local/abcarian/la-me-abcarian-montecito-20180123-story.html

National Aeronautics and Space Administration. "Deadly Debris Flows in Montecito." https://earthobservatory.nasa.gov/IOTD/view.php?id=91573

National Broadcasting Company. "Scientists Voyage to the White Shark Café." http://www.nbcrightnow.com/story/38329615/scientists-voyage-to-the-white-shark-café

National Geographic. "Amazing Video: Inside the World's Largest Gathering of Snakes." https://news.nationalgeographic.com/news/2014/06/140626-snakes-narcisse-animals-mating-sex-animals-world/#close

National Geographic. "Kentucky Sinkhole Eats Corvettes, Raises Questions." https://news.nationalgeographic.com/news/2014/02/140213-corvette-sinkhole-kentucky-museum-science/

National Oceanic and Atmospheric Administration. "NOAA/USGS Demonstration Flash-Flood and Debris-Flow Early Warning System." https://www.wrh.noaa.gov/lox/main.php?suite=hydrology&page=debris-flow_project

National Oceanic and Atmospheric Administration. "Saildrone." https://www.pmel.noaa.gov/ocs/saildrone

National Oceanic and Atmospheric Administration. "What Is a Rogue Wave?" https://oceanservice.noaa.gov/facts/roguewaves.html

National Public Radio. "Great White Sharks Have a Secret 'Café,' and They Led Scientists Right to It." https://www.npr.org/sections /thetwo-way/2018/05/28/613394086/great-white-sharks-have-a -secret-cafe-and-they-led-scientists-right-to-it

New York Times. "A Freak-Wave Flip-Out? Not Likely." http://www .nytimes.com/2006/05/14/weekinreview/14basicB.html

New York Times. "A Mudslide, Foretold." https://www.nytimes.com/2014 /03/30/opinion/sunday/egan-at-home-when-the-earth-moves.html

New York Times. "Rogue Giants at Sea." http://www.nytimes.com/2006 /07/11/science/11wave.html

Orlando Sentinel. "Pictures: Winter Park Sinkhole." http://www .orlandosentinel.com/news/nationworld/os-fla360-pictures-winter -park-sinkhole-20121113-photogallery.html

Popular Science. "What the Heck Are Sinkholes Anyway?" https://www .popsci.com/what-are-sinkholes-anyway

Prager, Ellen. *Chasing Science at Sea: Racing Hurricanes, Stalking Sharks, and Living Undersea with Ocean Experts*. Chicago: University of Chicago Press, 2008.

Schmidt Ocean Institute. "Voyage to the White Shark Café." https:// schmidtocean.org/cruise/voyage-white-shark-cafe/

Scientific American. "Looming Landslide Stokes Fears, May Help Disaster Predictions." https://www.scientificamerican.com/article/looming -landslide-stokes-fears-may-help-disaster-predictions/

Scientific American. "The Real Sea Monsters: On the Hunt for Rogue Waves." https://www.scientificamerican.com/article/rogue-waves -ocean-energy-forecasting/

Shark Sider. "Do You Know What Is More Dangerous Than a Shark?" https://www.sharksider.com/know-dangerous-shark%E2%80%A8/

Sidder, A. "Tracking Landslide Hazards around the World, Pixel by Pixel." *Eos* 97 (2016). https://doi.org/10.1029/2016EO060583

Stanford University. "Electronic Tracking System Allows Scientists to Tail White Sharks More Effectively." https://news.stanford.edu /news/2008/march5/sharkssr-030508.html

Surfline. "La Jolla Shores Rip Current Incident—Explained by Sean Collins." http://www.surfline.com/surf-science/la-jolla-shores-rip -current-incident---explained-by-sean-collins---forecaster-blog _57986/

Sydney Morning Herald. "Seattle Mudslide: 1999 Report Warned of 'Catastrophic Failure' at Site." https://www.smh.com.au/world/seattle

-mudslide-1999-report-warned-of-catastrophic-failure-at-site
-20140325-zqmuc.html

United States Geological Survey. "The Science of Sinkholes." https://
www2.usgs.gov/blogs/features/usgs_top_story/the-science-of
-sinkholes/

United States Geological Survey. "USGS Geologists Join Efforts in
Montecito to Assess Debris-Flow Aftermath." https://www.usgs.gov
/news/usgs-geologists-join-efforts-montecito-assess-debris-flow
-aftermath

United States Lifesaving Association. "Rip Currents." http://www.usla
.org/?page=RIPCURRENTS

University of South Florida. "Cars and Karst: Investigating the National
Corvette Museum Sinkhole." http://scholarcommons.usf.edu/sink
hole_2015/ProceedingswithProgram/Mgmt_Regs_Education/6/

USA Today. "Watch Sinkhole Swallow 8 Corvettes at Museum." https://
www.usatoday.com/story/news/nation/2014/02/12/corvette
-museum-sinkhole/5417171/

Washington Post. "Before and after the Mudslides in Montecito." https://
www.washingtonpost.com/graphics/2018/national/montecito
-before-after/

Washington Post. "Everything You Need to Know about the Washington
Landslide." https://www.washingtonpost.com/news/post-nation/wp
/2014/03/24/everything-you-need-to-know-about-the-washington
-landslide/

Wired. "How Mudslide Becomes a Deadly Tsunami of Rocks and Sludge."
https://www.wired.com/story/post-wildfire-mudslide/